A Night to Remember

WALTER LORD

With a Foreword by Julian Fellowes
and an Introduction by Brian Lavery

PENGUIN BOOKS

PENGUIN BOOKS

Published by the Penguin Group
Penguin Books Ltd, 80 Strand, London WC2R 0RL, England
Penguin Group (USA) Inc., 375 Hudson Street, New York, New York 10014, USA
Penguin Group (Canada), 90 Eglinton Avenue East, Suite 700, Toronto, Ontario, Canada M4P 2Y3
(a division of Pearson Penguin Canada Inc.)
Penguin Ireland, 25 St Stephen's Green, Dublin 2, Ireland (a division of Penguin Books Ltd)
Penguin Group (Australia), 250 Camberwell Road, Camberwell, Victoria 3124, Australia
(a division of Pearson Australia Group Pty Ltd)
Penguin Books India Pvt Ltd, 11 Community Centre, Panchsheel Park, New Delhi – 110 017, India
Penguin Group (NZ), 67 Apollo Drive, Rosedale, Auckland 0632, New Zealand
(a division of Pearson New Zealand Ltd)
Penguin Books (South Africa) (Pty) Ltd, Block D, Rosebank Office Park, 181 Jan Smuts Avenue,
Parktown North, Gauteng 2193, South Africa

Penguin Books Ltd, Registered Offices: 80 Strand, London WC2R 0RL, England

www.penguin.com

First published by Longmans Green & Co. 1956
Large-format revised and illustrated edition first published by Allen Lane 1976
Published in Penguin Books 1978
Reissued in this edition with a new Foreword and a new Introduction in Penguin Books 2012
003

Set in 12.5/14.75 pt Garamond MT Std
Typeset by Palimpsest Book Production Limited, Falkirk, Stirlingshire
Printed in Great Britain by Clays Ltd, St Ives plc

ISBN: 978-0-141-39969-0

www.greenpenguin.co.uk

PENGUIN BOOKS

A Night to Remember

'Absolutely gripping and unputdownable'
David McCullough, Pulitzer prize-winning author

'Walter Lord singlehandedly revived interest in the *Titanic* . . . an electrifying book'
John Maxtone-Graham, maritime historian and author

'*A Night to Remember* was a new kind of narrative history – quick, episodic, unsolemn. Its immense success inspired a film of the same name three years later'
Ian Jack, *Guardian*

'Devotion, gallantry . . . Benjamin Guggenheim changing to evening clothes to meet death; Mrs Isador Straus clinging to her husband, refusing to get in a lifeboat; Arthur Ryerson giving his lifebelt to his wife's maid . . . A book to remember'
Chicago Tribune

'Seamless and skilful . . . it's clear why this is many a researcher's *Titanic* bible'
Entertainment Weekly

'Enthralling from the first word to the last'
Atlantic Monthly

ABOUT THE AUTHORS

A graduate of Princeton University and Yale Law, Walter Lord served in England with the American Intelligence Service during the Second World War. His interest in the *Titanic* dates back to 1926 when, at ten years old, he persuaded his family to cross the Atlantic on the *Olympic*, sister ship to the doomed ocean liner. Lord was renowned for his knowledge of the *Titanic* catastrophe, serving as consultant to director James Cameron during the filming of *Titanic*. *A Night to Remember* was published in 1956 and has never been out of print. Walter Lord died in 2002.

Julian Fellowes is an actor, writer, director and producer. His film and television work includes *Gosford Park*, *Downton Abbey* and *Titanic*. His novels include *Snobs* and *Past Imperfect*.

Brian Lavery is Curator Emeritus at the National Maritime Museum, Greenwich. He is the author of books including *Ship: The Epic Story of Maritime Adventure*. He was consultant on the film *Master and Commander* and the BBC's *Empire of the Seas*.

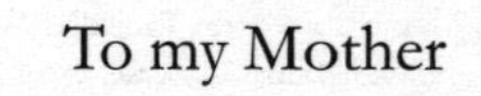
To my Mother

Contents

Foreword by Julian Fellowes

There are certain episodes in the past which fix like a burr on our imaginations, events in history which will not let us go. They are generally tragic ones: the destruction of Pompeii, the plague and fire in the London of the 1660s, the French Revolution. But few of these outrank that single incident, just a century ago, when a luxury liner, the very acme of its own type and time, struck an iceberg in the North Atlantic at 11.40 on the night of 14 April 1912, and sank just over two and a half hours later, thereby giving birth to books and films and memoirs and articles without number.

It is hard to pin down exactly why this tragedy still haunts us to the degree that it does, when the last of the infant passengers to survive have now gone to their reward. Maybe it is because the ship seemed, even then, to represent that proud, pre-war world in miniature, from the industrialists and peeresses and millionaires and Broadway producers who sat about the vast staterooms in first class, to the Irish and German and Scandinavian immigrants packed into third, carrying with them all they possessed, on their way to a new life in America.

There were the passengers in second class, too, professionals and their wives, and salesmen with samples of wares or order books at the ready, all set to make a deal with the entrepreneurs of the New World. And there was the crew, the boilermen and deckhands, the stewards and stewardesses, and, of course, the officers, who would find themselves at the centre of the drama of the ship's final hours. And as they headed for destruction, so did the larger world they represented, which would soon hit its own iceberg in the shape of the First World War.

Walter Lord begins his account of the disaster with a curious fact: in 1898 one Morgan Robertson wrote a novel about a fabulous liner, packed with the rich and fashionable folk of the day, which crashed into an iceberg and sank. The book was called *Futility* and the events predicted in it would become startlingly true. It seems to have been the discovery of this eerie coincidence that inspired Lord to take on the mantle of Chief Chronicler of the *Titanic*.

He would have many imitators, but what continues to mark his version apart from the rest is its extraordinary economy. He manages to convey both the detail and the sweep, the little sorrows and the all-embracing horror, in prose which is minutely researched but never dense. His style is serious, moving and, above all, readable. In my own investigation into the truth behind the sinking, I never came across another book to rival it.

The *Titanic* has spawned its own legends, its own heroes and heroines, but, as so often in life, the truth is a little more complicated. Some of these stars, the famously 'unsinkable' Molly Brown, for instance, or John Hart the third-class steward, or the stalwart Countess of Rothes, prove satisfactorily authentic when they are researched. Mrs Brown did indeed take the oars and try to get the lifeboats to go back for survivors; John Hart did lead parties up from steerage to the boat deck on his own initiative and he did get them away to safety; Lady Rothes did take over the tiller, and corresponded with the sailor in charge of her boat for the rest of their lives.

But then it was Charles Lightoller, second officer, one of the accepted heroes of the sinking, who decided not to fill the boats to capacity, and to take 'women and children only' (rather than the more usual 'women and children first'), his idea being that the men could swim out to join their womenfolk once the boats were safely launched. This doomed plan seems to have been arrived at because Lightoller was unaware that the boats had been tested full in Belfast, and failed to recognize that, after a short time, the hatches from which the men were to swim would be unreachable or that the water was too cold to survive in for more than a few minutes. As it was, the boats rowed away from the wreck as soon as they touched the surface of the sea to escape the suction which never in fact happened.

So while Lightoller definitely was a very brave man and a real hero, his split-second decision not to take men and not to fill all the boats cost hundreds of lives.

I wonder, too, whether some of the villains have been justly treated. History has not been kind to the Duff Gordons, but the charge against them of paying the sailors to keep away from the drowning was never proved. If they *were* afraid to return for fear of being swamped, it was no more than the fear felt in almost every boat.

And the Managing Director of the White Star Line,

Bruce Ismay, has had a hard press when he did not, as one often reads, get into the 'first' lifeboat to leave the ship. In fact, he climbed into the very last boat of any description, one of only two collapsibles to be successfully launched, to get away before the ship went down. Nor is there any solid evidence that he was responsible for the increase in speed, since it was White Star's clear and stated policy that they sold luxury rather than a record crossing. It is not anyway realistic to exonerate Captain Smith from the decision to go faster, as some have tried to do, when the order could not have been carried out, no matter where it came from, without his approval. During those frightful last two hours, Ismay had in fact spent a good deal of the time helping women and children into the boats before the temptation to survive proved too much for him. I wonder if his subsequent title, The Coward of the *Titanic*, which cast such a shadow over the remainder of his life, was quite merited.

I was recently in Budapest, where they were filming my scripts for the ITV/Indigo production of the story. Standing alone on the huge sets, astonishing replicas of the promenade deck and the boat deck, it was impossible not to think of that moment, a hundred years before, when some of the great names of Belgravia and Newport stood, in silent and dignified groups, waiting to learn their fate. The American Croesus John Jacob Astor and his pregnant young wife, Madeleine; the banking Wideners of Philadelphia; the railway king Charles Hays; the hedonist Benjamin Guggenheim; the silent-movie queen Dorothy Gibson; and behind them all those other men, women and children, rich and poor, old and young, from every background under the sun, for whom the next hundred minutes would deliver them either to life or to death.

Despite the wealth of new evidence gleaned from the discovery of the wreck, long after this book was first published, some of the mysteries of the sinking will probably never be solved. Why some piece of crucial equipment was mislaid, why this telegram was ignored, why that warning went unremarked.

And, like most of us, I am not sure of the lessons we can draw from this awful story; maybe just that we cannot know what Fate has in store, that we should not forget man is never the superior of nature, or simply that ordinary men and women are capable of acts of courage and kindness that make them great in the doing. Perhaps that's it. That savage events can inspire people to greatness.

Certainly, we cannot predict how we would behave in such a case, but we can hope and even pray that we would act as nobly as so many of the victims did, on that dark and terrible Atlantic night.

Julian Fellowes
August 2011

Introduction by Brian Lavery

When *A Night to Remember* was first published in the United States in 1955, Burke Wilkinson, in the *New York Times*, wrote that 'the author's style is so simple as to be almost an absence of style. But his great story needs no gilding, and he has given us that rarest of experiences – a book whose total effect is greater than the sum of its parts'. Stanley Walker of the *New York Herald Tribune* claimed that it was based on 'a kind of literary pointillism, the arrangement of contrasting bits of fact and emotion in such a fashion that a vividly real impression of an event is conveyed to the reader'.

When it was published in Britain in the following year, the reviewers were divided along political lines. In the *Illustrated London News* Sir John Squire, a poet and historian who had flirted with both Marxism and fascism in his time, found that Lord's populist approach to disaster 'slightly disgusts me'. The conservative *Times* thought that Lord had been unfair to the ship's owner, Bruce Ismay, who had escaped from the disaster. The high loss of life among the poor steerage passengers, it was claimed, was due to shortage of lifeboats and not class distinction. To the left-wing *New Statesman*, the disaster was caused by 'complacency and commercialism . . . the attempt of the White Star Line to wring the last penny out of the profitable Atlantic trade'.

But most reviewers saw it as having all the elements of a Greek tragedy.

Whatever the reviewers might think, the book sold very well and made Lord's reputation as a storyteller. It was filmed in 1958 with the highly popular British star Kenneth More in the role of Second Officer Charles Lightoller. The book helped to establish the idea of reporting a dramatic event through the accounts of ordinary people involved, which was used, for example, by Cornelius Ryan in *The Longest Day*. And it put the ageing story of the *Titanic* back in the forefront of the public consciousness.

However much one would like to say about the millions of people who built ships or sailed in them as passengers and crew, it is impossible for a maritime historian to escape from iconic characters such as Lord Nelson, and dramatic events such as the sinking of the *Titanic*. But it is quite likely that the *Titanic* would be almost forgotten now, or known only to specialists, if Walter Lord had not researched and published his most famous book at just the right moment.

By the 1950s the sinking had been overshadowed by two world wars, and it was no longer the greatest maritime disaster of all time – for Britain that distinction went to the *Lancastria*, sunk off Le Havre in 1940 with 2,500 people on board. In world terms the greatest loss of life was in the German *Wilhelm Gustloff* in the Baltic in 1945, when an estimated 7,000 people, including many refugees, were killed. But, of course, these were wartime disasters, unlike the *Titanic*, which sank in the calm waters of a peaceful world.

Lord was motivated by his love of the great liners, which he had travelled in as a boy with his parents, including a trip in the *Titanic*'s surviving sister-ship, *Olympic*, in 1927. He was fascinated by the idea of a closed society like a town afloat, even if the passengers were only on board for a week or so. He was researching his book at the right time, partly because many of the survivors were still alive and had fresh memories of events more than forty years before. Perhaps they were far enough from the disaster to get over any survivor's guilt, or the traumas of the night in question.

But Lord did not make use of one new invention which a modern researcher would regard as essential: the tape recorder. Nor did he take notes during the interviews, for fear of intimidating the witnesses. Instead, he prepared his questions for each interview very carefully, and memorized what was said, writing them down afterwards as soon as he found privacy.

The book was also published at exactly the right time. The television age was just beginning, but the public was already used to the immediacy of newsreel and radio reporting, and the highlighting of individual stories in the midst of historic events. Despite the reactions of some traditional historians, history was no longer about kings, queens and presidents but about how it was shaped by people of both high and low status.

Like most history books, *A Night to Remember* is about the time in which it was written as well as the period it describes. America in the 1950s was more prosperous than it had ever been, and it felt a great moral superiority after defeating the Nazis and taking on the Soviets in the Cold War. As Lord is careful to point out, it was far more classless than the society of 1912. Yet it too had a huge threat hanging over it, as the Soviets built up an increasingly terrifying nuclear arsenal, with thermonuclear bombs and ballistic missiles. Britain was no less threatened by the bomb, and its people had far less space to hide from it. It was about to face its own sinking moment, when the Suez Crisis of 1956 signalled the end of the British Empire. Lord does not deal with the issue of race, which was about to engulf the United States and, to a lesser extent, Britain. Many British shipping lines employed Africans and Asians as firemen, stewards and seamen, but not White Star. Almost everyone aboard the *Titanic*, both passengers and crew, was white (though there is casual mention of Chinese and Japanese) and racialism, which was an essential and largely unspoken feature of 1912 society, was directed against what were considered the 'lesser' European races such as the Italians.

Lord begins his story with the first sighting of the iceberg, and the world outside the ship appears only incidentally, increasing the feeling of peril and claustrophobia among those on board. He portrays the sinking as a slow-motion disaster, with its extent dawning on crew and passengers only by degrees. As he wrote in 1987, part of the appeal of the story relies on 'the initial refusal to believe that anything was wrong – card games continued in the smoking room; playful soccer matches on deck with chunks of ice broken off from the berg. Then the gradual dawning that there is real danger – the growing tilt of the deck, the rockets going off. And finally, the realization that the end is at hand, with no apparent escape.' Lord also starts with different levels in the ship – the lookout at the head of the mast, the officers on the bridge, the passengers in the saloons and the firemen down in the engine room. He explores the alternative hierarchies on board – the normal social one and the sea discipline, with officers commanding seamen, who in turn are nominally in charge of the passengers in a lifeboat – though in real life they were often challenged successfully by the first-class passengers, who believed they had a right to rule in any circumstances.

When he mentions it at all, Lord portrays the world outside the *Titanic* as a very stable one, only ended in later years by war and taxation, but it is worth remembering that Britain was in turmoil at the time, with militant suffragettes, very bitter strikes and Ireland on the verge of rebellion. Nevertheless, Lord convinces us that the social order was maintained on board the ship, as stewards and valets helped their masters prepare for the lifeboats. And the sacrifice of third-class passengers went largely unchallenged by the inquiries in Britain and the United States. Lord tells many individual stories of heroism and cowardice, selfishness and generosity. The band did indeed play on, though not, apparently, 'Nearer My God to Thee'. Lord tells how Bruce Ismay bullied his way into a lifeboat, only to live the next twenty-five years in loneliness and shame.

The interaction between Europe and America had been one of the most dynamic factors in world history for four hundred years, and Lord's story taps into that, in particular the links between Britain and the United States, with a common language. They had fought together undefeated in the recent war, and one was in the process of handing over the mantle of world power with a generally liberal reputation to the other. Practically all white Americans had ancestors who had emigrated in ships like the *Titanic*, and millions more had crossed the Atlantic both ways in wartime. Yet the era of the great liner was about to end. The Boeing 707 began its service in 1957 and for the first time it was more economical to cross by air. The liner had a slow death paralleling the *Titanic* herself, but regular scheduled transatlantic services had ended by 1973.

Today just as many people take to the seas in cruise ships but somehow they do not have the same mythology. Passengers' motives might range from pure hedonism to intellectual discovery, but even at its best the modern cruise does not have the sense of purpose of the great liner bridging the old world and the new, the isolation of those aboard in a closed society for days at a time, the stark divisions by class. And, of course, voyaging today is much safer than it was a hundred years ago. Loss of life in the *Costa Concordia* disaster of 2012 was mercifully small, but even so there are echoes of the *Titanic* – scramble for lifeboats, apparent neglect of duty by those in authority and heroism by others.

The *Titanic* disaster was soon overtaken by far greater catastrophes as Europe moved into the First World War two years later – a technologically advanced, arrogant and class-ridden society steamed boldly into danger despite numerous warning signals. The public never forgot the

Titanic, but thirty years of destruction and savagery seemed to overshadow it. For ten years after 1945 there was a great flood of war memoirs and novels, often made into highly successful movies such as *The Cruel Sea* and *The Caine Mutiny*.

There was a certain amount of reaction by the mid-fifties. Peacetime conscription in Britain and the United States had created a generation ready to laugh at all things military, as reflected in films such as *Private's Progress* and television characters like Sergeant Bilko, which showed soldiers as essentially lazy and corrupt. The public was ready for new heroes and legends, or for older ones to be revived. Even so, as the *New York Times* reviewer commented in 1955, there were already fifty books on the *Titanic* disaster in the Library of Congress, including four novels and six books of verse. But none of them matched the immediacy and impact of Lord's work.

The *Titanic* legend had another enormous boost in 1985 when Dr Robert Ballard announced his discovery of the wreck two miles under the Atlantic. Lord was sceptical about this when it happened. 'At first I thought that the discovery might spoil some of the allure. Part of the spell seemed to depend on the great ship, still hauntingly beautiful in her final moments, disappearing beneath the sea forever. But soon it became clear that the discovery actually added to the mystique.' The salvage revealed much about the technical details of the sinking, including the fact that the ship had broken in two close to the surface, and that the funnels had become detached one by one. It produced a great range of personal goods, salvaged by RMS Titanic Inc. These added little to the historical account, but they provided poignant and often emotional links with the past when they were shown in well-attended exhibitions around the world.

The third great *Titanic* revival came with James Cameron's film of 1997. Since the publication of *A Night to Remember*, the ship has been represented in fiction far more than any other in history. Offstage, it is included in almost every novel set in the period when it is necessary to get rid of a character or two – most recently, the loss of family heirs was the starting point for the highly successful TV series *Downton Abbey*. If all the fictional characters on board the ship could be counted, they would far outnumber the real passengers and crew. If they had any weight, they might sink the vessel without any help from the iceberg. Cameron's film relied heavily on Lord's research for many of the incidents described, a tribute to the merit of his work. The hero Jack Dawson finds it a little too easy to cross from third class to first or to go to the extreme bow, where only the crew were allowed, but in general the film was quite accurate.

The *Titanic* story remains as a legend. Though Walter Lord did everything humanly possible to find the absolute truth, he always knew that it could never be achieved: 'The best that can be done is to weigh the evidence carefully and give an honest opinion.' This is true of all historic research, and it is a tribute to Walter Lord's skill and honesty that his work is still influential and worthy of a reprint after more than half a century.

Brian Lavery, National Maritime Museum, Greenwich
January 2012

Preface

In 1898 a struggling author named Morgan Robertson concocted a novel about a fabulous Atlantic liner, far larger than any that had ever been built. Robertson loaded his ship with rich and complacent people and then wrecked it one cold April night on an iceberg. This somehow showed the futility of everything and, in fact, the book was called *Futility* when it appeared that year, published by the firm of M. F. Mansfield.

Fourteen years later a British shipping company named the White Star Line built a steamer remarkably like the one in Robertson's novel. The new liner was 66,000 tons displacement; Robertson's was 70,000 tons. The real ship was 882.5 feet long; the fictional one was 800 feet. Both vessels were triple screw and could make 24–5 knots. Both could carry about 3,000 people, and both had enough lifeboats for only a fraction of this number. But then, this didn't seem to matter because both were labelled 'unsinkable'.

On 10 April 1912 the real ship left Southampton on her maiden voyage to New York. Her cargo included a priceless copy of *The Rubáiyát* of Omar Khayyám and a list of passengers collectively worth 250 million dollars. On her way over she too struck an iceberg and went down on a cold April night.

Robertson called his ship the *Titan*; the White Star Line called its ship the *Titanic*. This is the story of her last night.

1. 'Another Belfast Trip'

High in the crow's-nest of the new White Star liner *Titanic*, lookout Frederick Fleet peered into a dazzling night. It was calm, clear and bitterly cold. There was no moon, but the cloudless sky blazed with stars. The Atlantic was like polished glass; people later said they had never seen it so smooth.

This was the fifth night of the *Titanic*'s maiden voyage to New York, and it was already clear that she was not only the largest but also the most glamorous ship in the world. Even the passengers' dogs were glamorous. John Jacob Astor had brought his Airedale Kitty. Henry Sleeper Harper, of the publishing family, had his prize Pekingese Sun Yat-Sen. Robert W. Daniel, the Philadelphia banker, was bringing back a champion French bulldog just purchased in Britain. Clarence Moore of Washington also had been dog-shopping, but the fifty pairs of English foxhounds he had bought for the Loudoun Hunt weren't making the trip.

That was all another world to Frederick Fleet. He was one of six lookouts carried by the *Titanic*, and the lookouts didn't worry about passenger problems. They were the 'eyes of the ship', and on this particular night Fleet had been warned to watch especially for icebergs.

So far, so good. On duty at 10 o'clock . . . a few words

about the ice problem with lookout Reginald Lee, who shared the same watch . . . a few more words about the cold . . . but mostly just silence, as the two men stared into the darkness.

Now the watch was almost over, and still there was nothing unusual. Just the night, the stars, the biting cold, the wind that whistled through the rigging as the *Titanic* raced across the calm, black sea at 22.5 knots. It was almost 11.40 p.m. on Sunday 14 April 1912.

Suddenly Fleet saw something directly ahead, even darker than the darkness. At first it was small (about the size, he thought, of two tables put together), but every second it grew larger and closer. Quickly Fleet banged the crow's-nest bell three times, the warning of danger ahead. At the same time he lifted the phone and rang the bridge.

'What did you see?' asked a calm voice at the other end.

'Iceberg right ahead,' replied Fleet.

'Thank you,' acknowledged the voice with curiously detached courtesy. Nothing more was said.

For the next thirty-seven seconds Fleet and Lee stood quietly side by side, watching the ice draw nearer. Now they were almost on top of it, and still the ship didn't turn. The berg towered wet and glistening far above the forecastle deck, and both men braced themselves for a crash. Then, miraculously, the bow began to swing to port. At the last second the stem shot into the clear, and the ice glided swiftly by along the starboard side. It looked to Fleet like a close shave.

At this moment Quartermaster George Thomas Rowe was standing watch on the after bridge. For him too, it had been an uneventful night – just the sea, the stars, the biting cold. As he paced the deck, he noticed what he and his mates called 'whiskers 'round the light' – tiny splinters of ice in the air, fine as dust, that gave off myriads of bright colours whenever caught in the glow of the deck lights.

Then suddenly he felt a curious motion break the steady rhythm of the engines. It was a little like coming alongside a dock wall rather heavily. He glanced forward – and stared again. A windjammer, sails set, seemed to be passing along the starboard side. Then he realized it was an iceberg, towering perhaps a hundred feet above the water. The next instant it was gone, drifting astern into the dark.

Meanwhile, down below in the first-class dining-saloon on D deck, four other members of the *Titanic*'s crew were sitting round one of the tables. The last diner had long since departed, and now the big white Jacobean room was empty except for this single group. They were dining-saloon stewards, indulging in the time-honoured pastime of all stewards off duty – they were gossiping about their passengers.

Then, as they sat there talking, a faint grinding jar seemed to come from somewhere deep inside the ship. It was not much, but enough to break the conversation and rattle the silver that was set for breakfast next morning.

Steward James Johnson felt he knew just what it was. He recognized the kind of shudder a ship gives when she drops a propeller blade, and he knew this sort of mishap meant a trip back to the Harland & Wolff shipyard at Belfast – with plenty of free time to enjoy the hospitality of the port. Somebody near him agreed and sang out cheerfully, 'Another Belfast trip!'

In the galley just to the stern, chief night baker Walter Belford was making rolls for the following day. (The honour of baking fancy pastry was reserved for the day shift.) When the jolt came, it impressed Belford more strongly than steward Johnson – perhaps because a pan of new rolls clattered off the top of the oven and scattered about the floor.

The passengers in their cabins felt the jar too, and tried to connect it with something familiar. Marguerite Frolicher, a young Swiss girl accompanying her father on a business trip, woke up with a start. Half-asleep, she could think only of the little white lake ferries at Zurich making a sloppy landing. Softly she said to herself, 'Isn't it funny . . . we're landing!'

Major Arthur Godfrey Peuchen, starting to undress for the night, thought it was like a heavy wave striking the ship. Mrs J. Stuart White was sitting on the edge of her bed, just reaching to turn out the light, when the ship seemed to roll over 'a thousand marbles'. To Lady Cosmo Duff Gordon, waking up from the jolt, it seemed 'as though somebody had drawn a giant finger along the side of the ship'. Mrs John Jacob Astor thought it was some mishap in the kitchen.

It seemed stronger to some than to others. Mrs Albert Caldwell pictured a large dog that had a baby kitten in its mouth and was shaking it. Mrs Walter B. Stephenson recalled the first ominous jolt when she was in the San Francisco earthquake – then decided this wasn't that bad.

Mrs E. D. Appleton felt hardly any shock at all, but she noticed an unpleasant ripping sound . . . like someone tearing a long, long strip of calico.

The jar meant more to J. Bruce Ismay, managing director of the White Star Line, who, in a festive mood, was going along for a ride on the *Titanic*'s first trip. Ismay woke up with a start in his de luxe suite on B deck – he felt sure the ship had struck something, but he didn't know what.

Some of the passengers already knew the answer. Mr and Mrs George A. Harder, a young honeymoon couple down in cabin E-50, were still awake when they heard a dull thump. Then they felt the ship quiver, and there was 'a sort of rumbling, scraping noise' along the ship's side. Mr Harder hopped out of bed and ran to the porthole. As he looked through the glass, he saw a wall of ice glide by.

The same thing happened to James B. McGough, a Gimbels buyer from Philadelphia, except his experience was more disturbing. His porthole was open, and as the berg brushed by, chunks of ice fell into the cabin.

Like Mr McGough, most of the *Titanic*'s passengers were in bed when the jar came. On this quiet, cold Sunday night a snug bunk seemed about the best place to be. But a few shipboard diehards were still up. As usual, most were in the first-class smoking-room on A deck.

And as usual, it was a very mixed group. Around one table sat Archie Butt, President Taft's military aide; Clarence Moore, the travelling Master of Hounds; Harry Widener, son of the Philadelphia streetcar magnate; and William Carter, another Main Liner. They were winding up a small dinner given by Widener's father in honour of

Captain Edward J. Smith, the ship's commander. The Captain had left early, the ladies had been packed off to bed, and now the men were enjoying a final cigar before turning in too. The conversation wandered from politics to Clarence Moore's adventures in West Virginia, the time he helped to interview the old feuding mountaineer Anse Hatfield.

Buried in a nearby leather armchair, Spencer V. Silverthorne, a young buyer for Nugent's department store in St Louis, browsed through a new best-seller, *The Virginian*. Not far off, Lucien P. Smith (still another Philadelphian) struggled gamely through the linguistic problems of a bridge game with three Frenchmen.

At another table the ship's young set was enjoying a somewhat noisier game of bridge. Normally the young set preferred the livelier Café Parisien, just below on B deck, and at first tonight was no exception. But it grew so cold that around 11.30 the girls went off to bed, and the men strolled up to the smoking-room for a nightcap. Most of the group stuck to highballs; Hugh Woolner, son of the English sculptor, took a hot whisky and water; Lieutenant Hokan Bjornstrom Steffanson, a young Swedish military attaché on his way to Washington, chose a hot lemonade.

Somebody produced a deck of cards, and as they sat playing and laughing, suddenly there came that grinding jar. Not much of a shock, but enough to give a man a start – Mr Silverthorne still sits up with a jolt when he tells it. In an instant the smoking-room steward and Mr Silverthorne were on their feet . . . through the aft door . . .

past the Palm Court . . . and out on to the deck. They were just in time to see the iceberg scraping along the starboard side, a little higher than the boat deck. As it slid by, they watched chunks of ice breaking off and tumbling into the water. In another moment it faded into the darkness astern.

Others in the smoking-room were pouring out now. As Hugh Woolner reached the deck, he heard a man call out, 'We hit an iceberg – there it is!'

Woolner squinted into the night. About 150 yards astern he made out a mountain of ice standing black against the starlit sky. Then it vanished into the dark.

The excitement, too, soon disappeared. The *Titanic* seemed as solid as ever, and it was too bitterly cold to stay outside any longer. Slowly the group filed back. Woolner picked up his hand, and the bridge game went on. The last man inside thought, as he slammed the deck door, that the engines were stopping.

He was right. Up on the bridge First Officer William M. Murdoch had just pulled the engine-room telegraph handle all the way to 'Stop'. Murdoch was in charge of the bridge this watch, and it was his problem, once Fleet phoned the warning. A tense minute had passed since then – orders to Quartermaster Hitchens to turn the wheel hard-a-starboard . . . a yank on the engine-room telegraph for 'Full speed astern' . . . a hard push on the button closing the watertight doors . . . and finally those thirty-seven seconds of breathless waiting.

Now the waiting was over, and it was also clearly too late. As the grinding noise died away, Captain Smith rushed on to the bridge from his cabin next to the wheelhouse. There were a few quick words:

'Mr Murdoch, what was that?'

'An iceberg, sir. I hard-a-starboarded and reversed the engines, and I was going to hard-a-port around it, but she was too close. I couldn't do any more.'

'Close the emergency doors.'

'The doors are already closed.'

They were closed all right. Down in boiler room No. 6 fireman Fred Barrett had been talking to second engineer James Hesketh when the warning bell sounded and the light flashed red above the watertight door leading to the stern. A quick shout of warning – an ear-splitting crash – and the whole starboard side of the ship seemed to give way. The sea cascaded in, swirling about the pipes and valves, and the two men leaped through the door as it slammed down behind them.

Barrett found things almost as bad where he was now, in boiler room No. 5. The gash ran into No. 5 about two feet beyond the closed compartment door, and a fat jet of sea-water was spouting through the hole. Nearby, trimmer George Cavell was digging himself out of an avalanche of coal that had poured out of a bunker with the impact. Another stoker mournfully studied an overturned bowl of soup that had been warming on a piece of machinery.

It was dry in the other boiler rooms further aft, but the scene was pretty much the same – men picking themselves up, calling back and forth, asking what had happened. It was hard to figure out. Until now the *Titanic* had been a picnic. Being a new ship on her maiden voyage, everything was clean. She was, as fireman George Kemish still recalls, 'a good job . . . not what we were accustomed to in old ships, slogging our guts out and nearly roasted by the heat'.

All the firemen had to do was keep the furnaces full. No need to work the fires with slice bars, pricker bars and rakes. So on this Sunday night the men were taking it easy – sitting around on buckets and the trimmers' iron wheelbarrows, gossiping, waiting for the twelve-to-four watch to come on.

Then came that thud . . . the grinding, tearing sound . . . the telegraphs ringing wildly . . . the watertight doors crashing down. Most of the men couldn't imagine what it was – the story spread that the *Titanic* had gone aground just off the Banks of Newfoundland. Many of them still thought so, even after a trimmer came running down from above shouting, 'Blimey! We've struck an iceberg!'

About ten miles away Third Officer Charles Victor Groves stood on the bridge of the Leyland liner *Californian*, bound from London to Boston. A plodding 6,000-tonner, she had room for forty-seven passengers, but none were being carried just now. On this Sunday night she had been stopped since 10.30 p.m., completely blocked by drifting ice.

At about 11.10 Groves noticed the lights of another ship, racing up from the east on the starboard side. As the newcomer rapidly overhauled the motionless *Californian*, a blaze of deck lights showed she was a large passenger liner. Around 11.30 he knocked on the Venetian door of the chart room and told Captain Stanley Lord about it. Lord suggested contacting the new arrival by Morse lamp, and Groves prepared to do this.

Then, at about 11.40, he saw the big ship suddenly stop and put out most of her lights. This didn't surprise Groves very much. He had spent some time in the Far East trade, where they usually put deck lights out at midnight to encourage the passengers to turn in. It never occurred to him that perhaps the lights were still on . . . that they only seemed to go out because she was no longer broadside but had veered sharply to port.

2. 'There's Talk of an Iceberg, Ma'am'

Almost as if nothing had happened, lookout Fleet resumed his watch, Mrs Astor lay back in her bed, and Lieutenant Steffanson returned to his hot lemonade.

At the request of several passengers second-class smoking-room steward James Witter went off to investigate the jar. But two tables of card players hardly looked up. Normally the White Star Line allowed no card playing on Sunday, and tonight the passengers wanted to take full advantage of the chief steward's unexpected largesse.

There was no one in the second-class lounge to send the librarian looking, so he continued sitting at his table, quietly counting the day's loan slips.

Through the long white corridors that led to the staterooms came only the murmurs of people chatting in their cabins . . . the distant slam of some deck-pantry door . . . occasionally the click of unhurried high heels – all the usual sounds of a liner at night.

Everything seemed perfectly normal – yet not quite. In his cabin on B deck, seventeen-year-old Jack Thayer had just called good night to his father and mother, Mr and Mrs John B. Thayer of Philadelphia. The Thayers had connecting staterooms, an arrangement compatible with Mr Thayer's position as Second Vice-President of the Pennsylvania Railroad. Now, as young Jack stood buttoning his pyjama jacket, the steady hum of the breeze through his half-opened porthole suddenly stopped.

One deck below, Mr and Mrs Henry B. Harris sat in their cabin playing double canfield. Mr Harris, a Broadway producer, was dog-tired, and Mrs Harris had just broken her arm. There was little conversation as Mrs Harris idly watched her dresses sway on their hangers from the ship's vibration. Suddenly she noticed they had stopped jiggling.

Another deck below, Lawrence Beesley, a young science master at Dulwich College, lay in his second-class bunk reading, pleasantly lulled by the dancing motion of the mattress. Suddenly the mattress was still.

The creaking woodwork, the distant rhythm of the engines, the steady rattle of the glass dome over the A deck foyer – all the familiar shipboard sounds vanished as the *Titanic* glided to a stop. Far more than any jolt, silence stirred the passengers.

Steward bells began ringing, but it was hard to learn anything. 'Why have we stopped?' Lawrence Beesley asked a passing steward. 'I don't know, sir,' came a typical answer, 'but I don't suppose it's much.'

Mrs Arthur Ryerson, of the steel family, had somewhat better results. 'There's talk of an iceberg, ma'am,' exclaimed steward Bishop. 'And they have stopped, not to run over it.' While her French maid Victorine hovered in the background, Mrs Ryerson pondered what to do. Mr Ryerson was having his first good sleep since the start of the trip, and she hated to wake him. She walked over to the square, heavy glass window that opened directly on to the sea. Outside, she saw only a calm, beautiful night. She decided to let him sleep.

Others refused to let well enough alone. With the restless curiosity that afflicts everyone on board ship, some of the *Titanic*'s passengers began exploring for an answer.

In C-51 Colonel Archibald Gracie, an amateur military historian by way of West Point and an independent income, methodically donned underwear, long stockings, shoes, trousers, a Norfolk jacket, and then puffed up to the boat deck. Jack Thayer simply threw an overcoat over his pyjamas and took off, calling to his parents that he was 'going out to see the fun'.

On deck there was little fun to be seen; nor was there any sign of danger. For the most part the explorers wandered aimlessly about or stood by the rail, staring into the empty night for some clue to the trouble. The *Titanic* lay dead in the water, three of her four huge funnels blowing off steam with a roar that shattered the quiet, starlit night. Otherwise everything was normal. Towards the stern of the boat deck an elderly couple strolled arm in arm, oblivious of the roaring steam and the little knots of passengers roving about.

It was so bitterly cold, and there was so little to be seen, that most of the people came inside again. Entering the magnificent foyer on A deck, they found others who had risen but preferred to stay inside where it was warm.

Mingling together, they made a curious picture. Their dress was an odd mixture of bathrobes, evening clothes, fur coats, turtle-neck sweaters. The setting was equally incongruous – the huge glass dome overhead . . . the dignified oak panelling . . . the magnificent balustrades with their wrought-iron scrollwork . . . and, looking down on them all, an incredible wall clock adorned with two bronze nymphs, somehow symbolizing Honour and Glory crowning Time.

'Oh, it'll be a few hours and we'll be on the way again,' a steward vaguely explained to first-class passenger George Harder.

'Looks like we've lost a propeller, but it'll give us more time for bridge,' called Howard Case, the London manager of Vacuum Oil, to Fred Seward, a New York lawyer. Perhaps Mr Case got his theory from steward Johnson, still contemplating a sojourn in Belfast. In any event, most of the passengers had better information by this time.

'What do you think?' exclaimed Harvey Collyer to his wife, as he returned to their cabin from a tour around the deck. 'We've struck an iceberg – a big one – but there's no danger. An officer told me so!' The Collyers were travelling second class, on their way from Britain to a fruit farm just purchased in Fayette Valley, Idaho. They were novices on the Atlantic, and perhaps the news would have roused Mrs Collyer, but the dinner that night had been too rich. So she just asked her husband if anybody seemed frightened, and when he said no, she lay back again in her bunk.

John Jacob Astor seemed equally unperturbed. Returning to his suite after going up to investigate, he explained to Mrs Astor that the ship had struck ice, but it didn't look serious. He was very calm and Mrs Astor wasn't a bit alarmed.

'What do they say is the trouble?' asked William T.

Stead, a leading British spiritualist, reformer, evangelist, and editor, all rolled into one. A professional individualist, he seemed almost to have planned his arrival on deck later than the others.

'Icebergs,' briefly explained Frank Millet, the distinguished American painter.

'Well,' Stead shrugged, 'I guess it's nothing serious; I'm going back to my cabin to read.'

Mr and Mrs Dickinson Bishop of Dowagiac, Michigan, had the same reaction. When a deck steward assured them, 'We have only struck a little piece of ice and passed it,' the Bishops returned to their stateroom and undressed again. Mr Bishop picked up a book and started to read, but soon he was interrupted by a knock on the door. It was Mr Albert A. Stewart, an ebullient old gentleman who had a large interest in the Barnum and Bailey Circus: 'Come on out and amuse yourself!'

Others had the same idea. First-class passenger Peter Daly heard one young lady tell another, 'Oh, come and let's see the berg – we have never seen one before.'

And in the second-class smoking-room somebody facetiously asked whether he could get some ice from the berg for his highball.

He could indeed. When the *Titanic* brushed by, several tons of ice crumbled off the berg and landed on the starboard well deck, just opposite the foremast. This was third-class recreation space, and the ice was soon discovered by steerage passengers coming up to investigate. From her cabin window on B deck, Mrs Natalie Wick watched them playfully throwing chunks at each other.

The ice soon became quite a tourist attraction. Major Arthur Godfrey Peuchen, a middle-aged manufacturing chemist from Toronto, used the opportunity to descend on a more distinguished compatriot, Charles M. Hays, President of the Grand Trunk Railroad. 'Mr Hays!' he cried, 'Have you seen the ice?'

When Mr Hays said he hadn't Peuchen followed through – 'If you care to see it, I will take you up on deck and show it to you.' And so they went all the way forward on A deck and looked down at the mild horseplay below.

Possession of the ice didn't remain a third-class monopoly for long. As Colonel Gracie stood in the A deck foyer, he was tapped on the shoulder by Clinch Smith, a New York society figure whose experiences already included sitting at Stanford White's table the night White was shot by Harry K. Thaw. 'Would you like,' asked Smith, 'a souvenir to take back to New York?' And he opened his hand to show a small piece of ice, flat like a pocket watch.

The same collector's instinct gripped others. Able seaman John Poingdestre picked up a sliver and showed it around the crew's mess room. A steerage passenger presented Fourth Officer Boxhall with a chunk about the size of a small basin. As greaser Walter Hurst lay half awake, his father-in-law – who shared the same quarters – came in and tossed a lump of ice into Hurst's bunk. A man entered the stewards' quarters, displaying a piece about as big as a teacup, and told steward F. Dent Ray, 'There are tons of ice forward!'

'Ah, well,' Ray yawned, 'that will not hurt.' And he prepared to go back to sleep.

A little more curious, first-class steward Henry Samuel Etches – off duty at the time of the crash – walked forward along the alleyway on E deck to investigate, and ran into a third-class passenger walking the other way. Before Etches could say anything, the passenger – as though confronting Etches with irrefutable evidence about something in dispute – threw a block of ice on to the deck and shouted, 'Will you believe it *now*?'

Soon there was far more disturbing evidence that all was not as it should be. By 11.50 – ten minutes after the collision – strange things could be seen and heard in the first six of the *Titanic*'s sixteen watertight compartments.

Lamp trimmer Samuel Hemming, lying off duty in his bunk, heard a curious hissing sound coming from the fore-peak, the compartment closest to the bow. He jumped up, went as far forward as he could, and discovered that it was air escaping from the forepeak locker where the anchor chains were stowed. Far below, water was pouring in so fast that air rushed out under tremendous pressure.

In the next compartment aft, containing the firemen's quarters and cargo hatch No. 1, leading fireman Charles Hendrickson was also roused by a curious sound. But here it was not air – it was water. When he looked down the spiral staircase that led to the passageway connecting the firemen's quarters with the stokeholds, he saw green sea-water swirling around the foot of the grated, cast-iron steps.

Steerage passenger Carl Johnson had an even more disturbing experience in the third compartment aft. This contained the cheapest passenger accommodation – lowest in the ship and closest to the bow. As Johnson got up to see what was causing a mild commotion outside his cabin, water seeped in under the door and around his feet. He decided to dress, and by the time his clothes were on, the water was over his shoes. With a detached, almost clinical interest, he noticed that it seemed to be of very even depth all over the floor. Nearby, steerage passenger Daniel Buckley was a little slower to react, and when he finally jumped out of his bunk, he splashed into water up to his ankles.

Five postal clerks working in the fourth compartment were much wetter. The *Titanic*'s post office took up two deck levels – the mail was stacked, along with first-class luggage, on the orlop deck and was sorted just above on G deck. The two levels were connected by a wide iron companionway, which continued up to F deck and the rest of the ship. Within five minutes water was sloshing around the knees of the postal clerks, as they dragged 200 sacks of registered mail up the companionway to the drier sorting room.

They might have spared themselves the trouble – in another five minutes the water reached the top of the steps and was lapping on to G deck. The clerks now abandoned the mail room altogether, retreating further up the companionway to F deck.

At the top of the stairs they found a married couple peering down at them. Mr and Mrs Norman Campbell Chambers of New York had been attracted by the noise while returning to their cabin after a fruitless trip to the promenade deck. Now, the Chamberses and the postal clerks watched the scene together, joking about the soaked baggage and wondering what might be in the letters they could see floating around the abandoned mail room.

Others joined them briefly from time to time – Fourth Officer Boxhall . . . assistant second steward Wheat . . . once even Captain Smith. But at no point could the Chamberses bring themselves to believe that anything they saw was really dangerous.

The fifth watertight compartment from the bow contained boiler room No. 6. This was where fireman Barrett and second engineer Hesketh jumped through the watertight door as it slammed down after the collision. Others didn't make it and scrambled up the escape ladders that laced their way topside. A few hung on, and after a moment some of the others came down again.

Shouts of 'Shut the dampers!' and then 'Draw the fires!' came from somewhere. Fireman George Beauchamp worked at fever pitch as the sea flooded in from the bunker door and up through the floor plates. In five minutes it was waist deep – black and slick with grease from the machinery. The air was heavy with steam. Fireman Beauchamp never did see who shouted the welcome words, 'That will do!' He was too relieved to care as he scurried up the ladder for the last time.

Just to the stern, second engineer Hesketh, now on the dry side of the watertight door, struggled to get boiler room No. 5 back to normal. The sea still spouted through a two-foot gash near the closed door, but assistant engineers Harvey and Wilson had a pump going, and it was keeping ahead of the water.

For a few moments the stokers stood by, aimlessly watching the engineers rig the pumps; then the engine room phoned to send them to the boat deck. They trooped up the escape ladder, but the bridge ordered them down again, and for a while they milled around the working alleyway on E deck – halfway up, halfway down – caught in the bureaucracy of a huge ship and wondering what to do next.

Meanwhile the lights went out in boiler room No. 5. Engineer Harvey ordered fireman Barrett, who had stayed behind, to go aft to the engine room for lanterns. The connecting doors were all shut; so Barrett had to climb to the top of the escape ladder, cross over, and go down the other side. By the time he retraced his steps, the engineers had the lights on again and the lanterns weren't needed.

Next, Harvey told Barrett to shut down the boilers – the pressure, built up while the ship was at full steam, now lifted the safety valves and was blowing joints. Barrett scrambled back up the ladder and drafted fifteen or twenty of the stokers wandering around E deck. They all clattered down and began wetting the fires. It was back-breaking work, boxing up the boilers and putting on dampers to stop the steam from rising. Fireman Kemish still remembers it with feeling: 'We certainly had one hell of a time putting those fires out . . .'

Clouds of steam gushed through the boiler room as the men sweated away. But gradually order returned. The lights burned bright, the place was clear of water, and, in No. 5 at any rate, everything seemed under control. There was an air of cheerful confidence by the time word spread that the men on the twelve-to-four watch were dragging their beds to the recreation deck because their rooms were flooded. The men on the eight-to-twelve watch paused in their work, thought this was a huge joke, and had a good laugh.

Up on the bridge, Captain Smith tried to piece the picture together. No one was better equipped to do it. After thirty-eight years' service with White Star, he was more than just senior captain of the line; he was a bearded patriarch, worshipped by crew and passengers alike. They loved everything about him – especially his wonderful combination of firmness and urbanity. It was strikingly evident in the matter of cigars. 'Cigars,' says his daughter, 'were his pleasure. And one was allowed to be in the room only if one was absolutely still, so that the blue cloud over his head never moved.'

Captain Smith was a natural leader, and on reaching the wheelhouse after the crash, he paused only long enough to visit the starboard wing of the bridge to see if the iceberg was still in sight. First Officer Murdoch and Fourth Officer Boxhall trailed along, and for a moment the three officers merely stood peering into the darkness. Boxhall thought he saw a dark shape astern, but he wasn't sure.

From then on all was business. Captain Smith sent Boxhall on a fast inspection of the ship. In a few minutes he was back: he had been as far forward in the steerage as he could go, and there was no sign of damage. This was the last good news Captain Smith heard that night.

Still worried, Smith now told Boxhall, 'Go down and find the carpenter and get him to sound the ship.' Boxhall wasn't even down the bridge ladder when he bumped into carpenter J. Hutchinson rushing up. As Hutchinson elbowed his way by, he gasped, 'She's making water fast!'

Hard on the carpenter's heels came mail clerk Iago Smith. He too pushed on towards the bridge, blurting out as he passed, 'The mail hold is filling rapidly!'

Next to arrive was Bruce Ismay. He had pulled a suit over his pyjamas, put on his carpet slippers, and climbed to the bridge to find out whether anything was happening that the president of the line should know. Captain Smith broke the news about the iceberg. Ismay then asked, 'Do you think the ship is seriously damaged?' A pause, and the captain slowly answered, 'I'm afraid she is.'

They would know soon enough. A call had been sent for Thomas Andrews, managing director of Harland & Wolff shipyard. As the *Titanic*'s builder, Andrews was making the maiden voyage to iron out any kinks in the ship. If anybody could figure out the situation, here was the man.

He was indeed a remarkable figure. As builder, he of course knew every detail about the *Titanic*. But there was so much more to him than that. Nothing was too great or too small for his attention. He even seemed able to anticipate how the ship would react to any situation. He understood ships the way some men are supposed to understand horses.

And he understood equally well the people who run ships. They all came to Andrews with their problems. One night it might be First Officer Murdoch, worried because he had been superseded by Chief Officer Wilde. The next night it might be a couple of quarrelling stewardesses who looked to Andrews as a sort of supreme court. This very evening chief baker Charles Joughin had made him a special loaf of bread.

So far, Andrews' trip had been what might be expected. All day long he roamed the ship, taking volumes of notes. At 6.45 every evening he dressed for dinner, dining usually with old Dr O'Loughlin, the ship's surgeon, who also had a way with the stewardesses. And then back to his stateroom A-36, piled high with plans and charts and blueprints. There he would assemble his notes and work out his recommendations.

Tonight the problems were typical – trouble with the restaurant galley hot press . . . the colouring of the pebble dashing on the private promenade decks was too dark . . . too many screws on all the stateroom hat hooks. There was also the plan to change part of the writing room into two more staterooms. The writing room had originally been planned partly as a place where the ladies could retire after dinner. But this was the twentieth century, and the ladies just wouldn't retire. Clearly, a smaller room would do.

Completely absorbed, Andrews scarcely noticed the jar and stirred from his blueprints only when he got Captain Smith's message that he was needed on the bridge.

In a few minutes Andrews and the captain were making their own tour – down the crew's stairway to attract less attention . . . along the labyrinth of corridors far below . . . by the water surging into the mail room . . . past the squash court, where the sea now lapped against the foul line on the backboard.

Threading their way back to the bridge, they passed

through the A deck foyer, still thronged with passengers standing around. Everybody studied the two men's faces for some sign of good news or bad; nobody could detect any clue.

Some of the crew weren't so guarded. In D-60, when Mrs Henry Sleeper Harper asked Dr O'Loughlin to persuade her sick husband to stay in bed, the old doctor exclaimed, 'They tell me the trunks are floating around in the hold; you may as well go on deck.'

In C-91 a young governess named Elizabeth Shutes sat with her charge, nineteen-year-old Margaret Graham. Seeing an officer pass the cabin door, Miss Shutes asked him if there was any danger. He cheerfully said no, but then she overheard him further down the hall say, 'We can keep the water out for a while.'

Miss Shutes glanced at Margaret, who was uneasily nibbling at a chicken sandwich. Her hand shook so badly the chicken kept falling out of the bread.

No one was asking questions along the working alleyway on E deck. This broad corridor was the quickest way from one end of the ship to the other – the officers called it 'Park Lane', the crew 'Scotland Road'. Now it was crowded with pushing, shoving people. Some were stokers forced out of boiler room No. 6, but most were steerage passengers, slowly working their way aft, carrying boxes, bags and even trunks.

These people didn't need to be told there was trouble. To those berthed far below on the starboard side, the crash was no 'faint grinding jar'. It was a 'tremendous noise' that sent them tumbling out of bed.

Mrs Celiney Yasbeck – a bride of fifty days – ran out into the corridor with her husband. Instead of making the long hike to the deck, it was easier to look below for trouble. In their night clothes they walked along to a door leading down to the boiler rooms and peeked through. Engineers were struggling to make repairs and get the pumps going. The Yasbecks needed no second glance – they rushed back to their cabin to dress.

Far above on A deck, second-class passenger Lawrence Beesley noticed a curious thing. As he started below to check his cabin, he felt certain the stairs 'weren't quite right'. They seemed level, and yet his feet didn't fall where they should. Somehow they strayed forward off balance . . . as though the steps were tilted down towards the bow.

Major Peuchen noticed it too. As he stood with Mr Hays at the forward end of A deck, looking down at the steerage passengers playing soccer with the loose ice, he sensed a very slight tilt in the deck. 'Why, she is listing!' he cried to Hays. 'She should not do that! The water is perfectly calm and the boat has stopped.'

'Oh, I don't know,' Mr Hays replied placidly, 'you cannot sink this boat.'

Others also felt the downward slant, but it seemed tactless to mention the matter. In boiler room No. 5, fireman Barrett decided to say nothing to the engineers working on the pumps. Far above in the A deck foyer, Colonel Gracie and Clinch Smith had the same reaction. On the bridge the commutator showed the *Titanic* slightly down at the head and listing five degrees to starboard.

Nearby, Andrews and Captain Smith did some fast

figuring. Water in the fore-peak . . . No. 1 hold . . . No. 2 hold . . . mail room . . . boiler room No. 6 . . . boiler room No. 5. Water fourteen feet above keel level in the first ten minutes, everywhere except boiler room No. 5. Put together, the facts showed a 300-foot gash, with the first five compartments hopelessly flooded.

What did this mean? Andrews quietly explained. The *Titanic* could float with any two of her sixteen watertight compartments flooded. She could float with any three of her first five compartments flooded. She could even float with all of her first four compartments gone. But no matter how they sliced it, she could not float with all of her first five compartments full.

The bulkhead between the fifth and sixth compartments went only as high as E deck. If the first five compartments were flooded, the bow would sink so low that water in the fifth compartment must overflow into the sixth. When this was full, it would overflow into the seventh, and so on. It was a mathematical certainty, pure and simple. There was no way out.

But it was still a shock. After all, the *Titanic* was considered unsinkable. And not just in the travel brochures. The highly technical magazine *Shipbuilder* described her compartment system in a special edition in 1911, pointing out, 'The captain may, by simply moving an electric switch, instantly close the doors throughout and make the vessel practically unsinkable.'

Now all the switches were pulled, and Andrews said it made no difference.

It was hard to face, and especially hard for Captain Smith.

Over fifty-nine years old, he was retiring after this trip. Might even have done it sooner, but he traditionally took the White Star ships on their maiden voyages. Only six years before, when he brought over the brand-new *Adriatic*, he remarked: 'I cannot imagine any condition which would cause a ship to founder. I cannot conceive of any vital disaster happening to this vessel. Modern shipbuilding has gone beyond that.' Now he stood on the bridge of a liner twice as big – twice as safe – and the builder told him it couldn't float.

At 12.05 a.m. – twenty-five minutes after that bumping, grinding jar – Captain Smith ordered Chief Officer Wilde to uncover the boats . . . First Officer Murdoch to muster the passengers . . . Sixth Officer Moody to get out the list of boat assignments . . . Fourth Officer Boxhall to wake up Second Officer Lightoller and Third Officer Pitman. The captain himself then walked about twenty yards down the port side of the boat deck to the wireless shack.

Inside, first operator John George Phillips and second operator Harold Bride showed no sign that they realized what was happening. It had been a tough day. In 1912 wireless was still an erratic novelty; range was short, operators were inexperienced, and signals were hard to catch. There was a lot of relaying, a lot of repeats, and a lot of frivolous private traffic. Passengers were fascinated by the new miracle, couldn't resist the temptation of sending messages to friends back home or on other ships.

All this Sunday the messages had piled up. It was enough to fray the nerves of any man working a fourteen-hour day at thirty dollars a month, and Phillips was no exception. Evening came, and still the bottomless in-basket, still the petty interferences. Only an hour ago – just when he was at last in good contact with Cape Race – the *Californian* barged in with some message about icebergs. She was so close she almost blew his ears off. No wonder he had snapped back, 'Shut up, shut up! I am busy; I am working Cape Race!'

It was such a hard day that second operator Bride decided to relieve Phillips at midnight, even though he wasn't due until 2.00 a.m. He woke up at about 11.55, brushed by the green curtain separating the sleeping quarters from the 'office', and asked Phillips how he was getting along. Phillips said he had just finished the Cape Race traffic. Bride padded back to his berth and took off his pyjamas. Phillips called after him that he thought the ship had been damaged somehow and they'd have to go back to Belfast.

In a couple of minutes Bride was dressed and took over the headphones. Phillips was hardly behind the green curtain when Captain Smith appeared: 'We've struck an iceberg and I'm having an inspection made to see what it has done to us. You'd better get ready to send out a call for assistance, but don't send it until I tell you.'

Then he left but returned again in a few minutes. This time he merely stuck his head in the doorway:

'Send the call for assistance.'

By now Phillips was back in the room. He asked the captain whether to use the regulation distress call. Smith replied, 'Yes, at once!'

He handed Phillips a slip of paper with the *Titanic*'s

position. Phillips took the headphones from Bride, and at 12.15 a.m. began tapping out the letters 'CQD' – at that time the usual international call of distress – followed by 'MGY', the call letters of the *Titanic*. Again and again, six times over, the signal rasped out into the cold, blue Atlantic night.

Ten miles away, Third Officer Groves of the *Californian* sat on the bunk of wireless operator Cyril F. Evans. Groves was young, alert and always interested in what was going on in the world. After work he liked to drop by Evans' wireless shack and pick up the latest news. He even liked to fool with the set.

This was all right with Evans. There weren't many officers on third-rate liners interested in the outside world, much less the wireless telegraphy. In fact, there weren't any others on the *Californian*. So he used to welcome Groves' visits.

But not tonight. It had been a hard day, and there was no operator to relieve him. Besides, he had been pretty roughly handled around 11.00 when he tried to break in on the *Titanic* and tell her about the ice blocking the *Californian*. So he lost no time tonight closing down his set at 11.30, his scheduled hour for going off duty. Now – dead tired – he was in no mood for chatting with anybody. Groves made a brave try: 'What ships have you got, Sparks?'

'Only the *Titanic*.' Evans scarcely bothered to glance up from his magazine.

Undeterred, Groves took the headphones and put them on. He was really getting quite good, if the message was simple enough. But he didn't know too much about the equipment. The *Californian*'s set had a magnetic detector that ran by clockwork. Groves didn't wind it up, and so he heard nothing.

Giving up, he put the phones back on the table, and went below to find livelier company. It was just a little after 12.15 a.m.

3. 'God Himself Could Not Sink This Ship'

The door to the cooks' quarters whacked open against the iron cot of assistant baker Charles Burgess. He woke up with a start and stared at second steward George Dodd standing in the doorway. Normally a rotund, jolly man, Dodd looked serious as he called, 'Get up, lads, we're sinking!'

Dodd moved forward to the waiters' quarters, where saloon steward William Moss was trying to rouse the men. Most of them were laughing and joking, when Dodd burst in, shouting, 'Get every man up! Don't let a man stay here!'

He moved on with Moss towards the stewards' quarters. Just outside, smoking-room steward Witter was already getting some disturbing news from carpenter Hutchinson: 'The bloody mail room is full.' Moss came up and added, 'It's really serious, Jim.'

The wisecracks that greeted the first warnings faded, and the crew tumbled out of their berths. Still half-asleep, baker Burgess pulled on pants, a shirt, no lifebelt. Walter Belford wore his white baker's coat, pants, didn't stop to put on his underdrawers. Steward Ray took more time; he wasn't worried – nevertheless he found himself putting on his shore suit. Steward Witter, already dressed, opened his trunk and filled his pockets with cigarettes . . . picked up the caul from his first child, which he always carried with him . . . then joined the crowd of men now swarming out into the working alleyway and up towards the boat stations.

Far forward, away from the uproar, trimmer Samuel Hemming climbed back into his bunk, satisfied that the hissing sound in the forepeak didn't mean very much. He was just drifting off to sleep when the ship's joiner leaned in, saying, 'If I were you, I'd turn out. She's making water one-two-three, and the racket court is getting filled up.' An instant later the boatswain appeared: 'Turn out, you fellows. You haven't half an hour to live. That is from Mr Andrews. Keep it to yourselves and let no one know.'

Certainly no one knew in the first-class smoking-room. The bridge game was going full blast again. Lieutenant Steffanson was still sipping his hot lemonade, and another hand was being dealt, when a ship's officer suddenly appeared at the door: 'Men, get on your lifebelts; there's trouble ahead.'

In her A deck stateroom, Mrs Washington Dodge lay in bed, waiting for Dr Dodge, Assessor for San Francisco, to dig up some news. The door opened and the doctor came in quietly: 'Ruth, the accident is rather a serious one; you had better come on deck at once.'

Two decks below, Mrs Lucien Smith – tired of waiting for Mr Smith to finish exploring – had gone back to sleep. Suddenly the lights snapped on, and she saw her husband standing by the bed, smiling down at her. Leisurely he explained, 'We are in the north and have struck an iceberg. It does not amount to anything but will probably delay us a day getting into New York. However, as a matter of form, the captain has ordered all ladies on deck.'

And so it went. No bells or sirens. No general alarm. But all over the *Titanic*, in one way or another, the word was passed.

It was very bewildering to eight-year-old Marshall Drew. When his aunt Mrs James Drew woke him and said she had to take him on deck, he sleepily protested he didn't want to get up. But Mrs Drew paid no attention.

It was no less bewildering to Major Arthur Peuchen, despite his sightseeing expedition to look at the ice. He heard the news on the grand staircase and could hardly believe it. Completely stunned, he stumbled to his cabin to change from evening dress into something warm.

For many, first word came from their stewards. John Hardy, second-class chief steward, personally roused twenty to twenty-four cabins. Each time he threw the door open wide, shouting, 'Everybody on deck with lifebelts on, at once!'

In first class it was more polite to knock. These were the days when a steward on a crack liner didn't have more than eight or nine cabins, and he was like a mother hen to all the passengers he served.

Steward Alfred Crawford was typical. He had spent thirty-one years handling difficult passengers, and now he knew just how to coax old Mr Albert Stewart into a life jacket. Then he stopped and tied the old gentleman's shoes.

In C-89, steward Andrew Cunningham helped William T. Stead into his lifebelt, while the great editor mildly complained that it was all a lot of nonsense. In B-84, steward Henry Samuel Etches worked like a solicitous tailor, fitting Benjamin Guggenheim for his lifebelt.

'This will hurt,' protested the mining and smelting king. Etches finally took the belt off altogether, made some adjustments, put it on again. Next, Guggenheim wanted to go on deck as he was, but Etches was adamant – it was much too cold. Ultimately Guggenheim submitted; Etches pulled a heavy sweater over him and sent him packing off topside.

Some of the passengers were even more difficult. At C-78, Etches found the door locked. When he knocked loudly with both hands, a man inside asked suspiciously, 'What is it?' and a woman added, 'Tell us what the trouble is.' Etches explained and again tried to get them to open the door. He had no luck, and after a few minutes' pleading he finally passed on to the next cabin.

In another part of the ship a locked door raised a different problem. It was jammed, and some passengers broke it down to release a man inside. At this point a steward arrived, threatening to have everybody arrested for damaging company property when the *Titanic* reached New York.

At 12.15 it was hard to know whether to joke or be serious – whether to chop down a door and be a hero, or chop it down and get arrested. No two people seemed to have the same reaction.

Mrs Arthur Ryerson felt there wasn't a moment to lose. She had long since abandoned the idea of letting Mr Ryerson sleep; now she scurried about trying to keep her family together. There were six to get ready – her husband, three children, governess and maid – and the children seemed so slow. Finally she gave up on her youngest daughter; just threw a fur coat over her nightgown and told her to come on.

There seemed all the time in the world to Mrs Lucien Smith. Slowly and with great care she dressed for whatever the night might bring – a heavy woollen dress, high shoes, two coats and a warm knitted hood. All the while Mr Smith chatted away about landing in New York, taking the train south, never mentioning the iceberg. As they started for the deck, Mrs Smith decided to go back for some jewellery. Here Mr Smith drew the line. He suggested it might be wiser not to bother with 'trifles'. As a compromise Mrs Smith picked up two favourite rings. Closing the door carefully behind them, the young couple headed up towards the boat deck.

Close behind came the Countess of Rothes and her cousin Gladys Cherry. They had difficulty putting on their lifebelts, and a passing gentleman paused to help them. He topped off the courtesy by handing them some raisins to eat.

The things people took with them showed how they felt. Adolf Dyker handed his wife a small satchel containing two gold watches, two diamond rings, a sapphire necklace and 200 Swedish crowns. Miss Edith Russell carried a musical toy pig (it played the 'Maxixe'). Stuart Collett, a young theological student travelling second class, took the Bible he promised his brother he'd always carry until they met again. Lawrence Beesley stuffed the pockets of his Norfolk jacket with the books he had been reading in bed. Norman Campbell Chambers pocketed a revolver and compass. Steward Johnson, by now anticipating far more than 'another Belfast trip', stuck four oranges under his blouse. Mrs Dickinson Bishop left behind 11,000 dollars in jewellery, then sent her husband back for her muff.

Major Arthur Peuchen looked at the tin box on the table in C-104. Inside were 200,000 dollars in bonds, 100,000 dollars in preferred stock. He thought a good deal about it as he took off his dinner jacket, put on two suits of long underwear and some heavy clothes.

Then he took a last look around the little cabin – the real brass bed . . . the green mesh net along the wall for valuables at night . . . the marble washstand . . . the wicker armchair . . . the horsehair sofa . . . the fan in the ceiling . . . the bells and electrical fixtures that on a liner always look as if they were installed as an afterthought.

Now his mind was made up. He slammed the door, leaving behind the tin box on the table. In another minute he was back. Quickly he picked up a good-luck pin and three oranges. As he left for the last time, the tin box was still on the table.

Out in the C deck foyer, Purser Herbert McElroy was urging everyone to stop standing around. As the Countess of Rothes passed, he called, 'Hurry, little lady, there is not much time. I'm glad you didn't ask me for your jewels as some ladies have.'

Into the halls they poured, gently prodded along by the crew. One room steward caught the eye of Miss Marguerite

Frolicher as she came down the corridor. Four days before, she had playfully teased him for putting a lifebelt in her stateroom, if the ship was meant to be so unsinkable. At the time he had laughed and assured her it was just a formality . . . she would never have to wear it. Remembering the exchange, he now smiled and reassured her, 'Don't be scared; it's all right.'

'I'm not scared,' she replied, 'I'm just seasick.'

Up the stairs they trooped – a hushed crowd in jumbled array. Under his overcoat Jack Thayer now sported a greenish tweed suit and vest, with another mohair vest underneath. Mr Robert Daniel, the Philadelphia banker, had on only woollen pyjamas. Mrs Turrell Cavendish wore a wrapper and Mr Cavendish's overcoat . . . Mrs John C. Hogeboom a fur coat over her nightgown . . . Mrs Ada Clark just a nightgown. Mrs Washington Dodge didn't bother to put on stockings under her high-button shoes, which flopped open because she didn't stop to button them. Mrs Astor looked right out of a bandbox in an attractive light dress, Mrs James J. Brown – a colourful Denver millionairess – equally stylish in a black velvet two-piece suit with black and white silk lapels.

Automobiling, as practised in 1912, affected the attire of many ladies – Mrs C. E. Henry Stengal wore a veil tightly pinned down over her floral hat, Madame de Villiers a long woollen motoring coat over her nightgown and evening slippers.

Young Alfred von Drachstedt, a twenty-year-old youth from Cologne, settled on a sweater and a pair of trousers, leaving behind a brand-new 2,133-dollar wardrobe that included walking sticks and a fountain pen, which he somehow felt was a special badge of distinction.

Second class was somewhat less elegantly disarrayed. Mr and Mrs Albert Caldwell – returning from Siam, where they taught at the Bangkok Christian College – had bought new clothes in London, but tonight they dressed in the oldest clothes they owned. Their baby Alden was wrapped in a blanket. Miss Elizabeth Nye wore a simple skirt, coat and slippers. Mrs Charlotte Collyer didn't bother to put up her hair, just tied it back with a ribbon. Her eight-year-old daughter Marjory had a steamer rug around her shoulders. Mr Collyer took little trouble dressing, because he expected to be back soon – he even left his watch lying on his pillow.

The scene in third class was particularly confusing because the White Star Line primly quartered the single men and single women at opposite ends of the *Titanic*. Now many of the men – who slept towards the bow – hurried aft to join the girls.

Katherine Gilnagh, a pert colleen not quite sixteen, heard a knock on the door. It was the young man who had caught her eye earlier that day playing the bagpipes on deck. He told her to get up – something was wrong with the ship. Anna Sjoblom, an eighteen-year-old Finnish girl bound for the Pacific Northwest, woke up when a young Danish swain came in to rouse her room-mate. He also gave Anna a lifebelt and urged her to come along. But she was too seasick to care. Eventually there was so much commotion that she went up after all, even though she still felt awful. She was quickly helped into a lifebelt by Alfred Wicklund, a schoolfriend from home.

Among these young men, Olaus Abelseth was especially worried. He was a twenty-six-year-old Norwegian heading for a South Dakota homestead, and an old family friend had put a sixteen-year-old daughter in his care until they reached Minneapolis. As he pushed his way aft along the E deck working alleyway, Minneapolis seemed a long way off.

Abelseth found the girl in the main steerage hallway on E deck. Then, along with his brother-in-law, a cousin and another girl, he climbed the broad, steep third-class stairs to the poop deck at the very stern of the ship.

Into the bitter night the whole crowd milled, each class automatically keeping to its own decks – first class in the centre of the ship, second a little aft, third at the very stern or in the well deck near the bow. Quietly they stood around waiting for the next orders . . . reasonably confident yet vaguely worried. With uneasy amusement they eyed how one another looked in lifebelts. There were a few half-hearted jokes.

'Well,' said Clinch Smith as a girl walked by carrying a Pomeranian, 'I suppose we ought to put a life preserver on the little doggie too.'

'Try this on,' a man told Mrs Vera Dick as he fastened on her life jacket. 'They are the very latest thing this season. Everybody is wearing them now.'

'They will keep you warm if you don't have to use them,' Captain Smith cheerfully explained to Mrs Alexander T. Compton of New Orleans.

At about 12.30 Colonel Gracie bumped into Fred Wright, the *Titanic*'s squash pro. Remembering he had reserved the court for 7.30 in the morning, Gracie tried a little joke of his own: 'Hadn't we better cancel that appointment?'

'Yes,' replied Wright. His voice was flat and without enthusiasm, but the wonder is he played along at all. He knew the water was now up to the squash-court ceiling.

In the brightly lit gym, just off the boat deck, Mr and Mrs Astor sat side by side on a pair of motionless mechanical horses. They wore their lifebelts, and Mr Astor had an extra one in his lap. He was slicing it open with his penknife, whiling away the time by showing his wife what was inside.

While the passengers joked and talked and waited, the crew moved swiftly to their stations. The boat teemed with seamen, stewards, firemen, chefs, ordered up from below.

A curiously late arrival was Fifth Officer Harold Godfrey Lowe. A tempestuous young Welshman, Lowe was hard to suppress. When he was fourteen, his father tried to apprentice him to a Liverpool businessman, but Lowe said he 'wouldn't work for nobody for nothing'. So he ran away to sea and a life after his own heart – schooners . . . square-riggers . . . five years steaming along the West African coast.

Now, at twenty-eight, he was making his first trip across the Atlantic. This Sunday night he was off duty and slept through the collision. Voices outside his cabin on the boat deck finally woke him up. When he looked out of the porthole and saw everybody in lifebelts, he catapulted out of bed, into his clothes, and rushed on deck to help. Not exactly an auspicious start, but then, as Lowe later explained to US senator Smith, 'You must remember that we do not have any too much sleep, and therefore when we sleep we die.'

Second Officer Charles Herbert Lightoller was late too, but for an entirely different reason. Like Lowe, he was off duty in his bunk when the *Titanic* hit, but he woke up instantly and, in his bare feet, ran out on the boat deck to see what was up. Nothing could be seen on either side of the ship, except on the starboard wing of the bridge, where he dimly made out Captain Smith and First Officer Murdoch. They too were peering out into the night.

Lightoller returned to his cabin and thought it over. Something undoubtedly was wrong with the ship – first that jar, now the silent engines. But he was *off* duty and, until called, it was no business of his. When they needed him, they would send for him. When this happened, he should be where they'd expect to find him. Lightoller got back into bed and lay awake waiting . . .

Five, fifteen, thirty minutes went by. He could now hear the roar of the funnels blowing off steam, the rising sound of voices, the clanking of gears. But still, his duty was to be where they'd expect to find him.

At 12.10 Fourth Officer Boxhall finally came bursting in: 'You know we have struck an iceberg.'

'I know we have struck something,' Lightoller replied, getting up and starting to dress.

'The water is up to F deck in the mail room,' continued Boxhall, by way of a little prodding. But no urging was needed. Lightoller was already well on the way. Cool, diligent, cautious, he knew his duty to the letter. He was the perfect Second Officer.

On the boat deck men began to clear the sixteen wooden lifeboats. There were eight on each side – a cluster of four towards the bow, then an open space of 190 feet, then another four towards the stern. Port boats had even numbers, starboard odd. They were numbered in sequence, starting from the bow. In addition, four canvas collapsible lifeboats – known as Englehardts – were stowed on deck. These could be fitted into the empty davits after the two forward boats were lowered. The collapsibles were lettered A, B, C and D.

All the boats together could carry 1,178 people. On this Sunday night there were 2,207 people on board the *Titanic*.

This mathematical discrepancy was known by none of the passengers and few of the crew, but most of them wouldn't have cared anyhow. The *Titanic* was unsinkable. Everybody said so. When Mrs Albert Caldwell was watching the deck hands carry up luggage at Southampton, she had asked one of them, 'Is this ship really non-sinkable?'

'Yes, lady,' he answered. 'God Himself could not sink this ship.'

So now the passengers stood calmly on the boat deck – unworried but very confused. There had been no boat drill. The passengers had no boat assignments. The crew had assignments, but hardly anybody bothered to look at the list. Now they were playing it strictly by ear – yet somehow the crew seemed to sense where they were needed and how to be useful. The years of discipline were paying off.

Little knots of men swarmed over each boat, taking off the canvas covers, clearing the masts and useless paraphernalia, putting in lanterns and tins of biscuits. Other men stood at the davits, fitting in cranks and uncoiling the lines. One by one the cranks were turned. The davits creaked, the pulleys squealed and the boats slowly swung out free of the ship. Next, a few feet of line were paid out, so that each boat would lie flush with the boat deck . . . or, in some cases, flush with promenade deck A directly below.

But the going was slow. Second Officer Lightoller, in charge of the port side, believed in channels, and Chief Officer Wilde's side seemed quite a bottleneck. When Lightoller asked permission to swing out, Wilde said, 'No, wait.' Lightoller finally went to the bridge and got orders direct from Captain Smith. Now Lightoller asked Wilde if he could load up. Again Wilde said no; again Lightoller went to the bridge; again Captain Smith gave him the nod: 'Yes, put the women and children in and lower away.'

Lightoller then lowered boat 4 level with A deck and ordered the women and children down to be loaded from there. It seemed safer that way – less chance of falling overboard, less distance to the water, and it helped clear the boat deck for hard work ahead. Too late he remembered the promenade deck was closed here and the windows were shut. While someone was sent to get the windows open, he hastily recalled everybody and moved aft to boat 6.

With one foot in No. 6 and one on deck, Lightoller now called for women and children. The response was anything but enthusiastic. Why trade the bright decks of the *Titanic* for a few dark hours in a rowboat? Even John Jacob Astor ridiculed the idea: 'We are safer here than in that little boat.'

As Mrs J. Stuart White climbed into No. 8, a friend called, 'When you get back you'll need a pass. You can't get back on tomorrow morning without a pass!'

When Mrs Constance Willard flatly refused to enter the boat, an exasperated officer finally shrugged: 'Don't waste time – let her go if she won't get in!'

And there was music to lull them too. Bandmaster Wallace Henry Hartley had assembled his men, and the band was playing ragtime. Just now they stood in the first-class lounge, where many of the passengers waited before orders came to lower the boats. Later they moved to the boat deck forward, near the entrance to the grand staircase. They looked a little nondescript – some in blue uniform coats, some in white jackets – but there was nothing wrong with their music.

Everything had been done to give the *Titanic* the best band on the Atlantic. The White Star Line had even raided the Cunarder *Mauretania* for bandmaster Hartley. Pianist Theodore Brailey and cellist Roger Bricoux were easily wooed from the *Carpathia*. 'Well, steward,' they happily told Robert Vaughan who served them on the little Cunarder, 'we will soon be on a decent ship with decent grub.' Bass player Fred Clark had never shipped before, but he was well known on the Scottish concert circuit, and the line brought him away too. First violinist Jock Hume hadn't yet played in any concerts, but his fiddle had a gay note the passengers seemed to love. And so it went – eight fine musicians who knew just what to do. Tonight the beat was fast, the music loud and cheerful.

On the starboard side things moved a little faster. But not fast enough for President Ismay, who dashed to and fro, urging the men to hurry. 'There's no time to lose!' he urged Third Officer Pitman, who was working on boat 5. Pitman shrugged him off – he didn't know Ismay and he had no time for an officious stranger in carpet slippers. Ismay told him to load the boat with women and children. This was too much for Pitman: 'I await the commander's orders,' he announced.

Suddenly it dawned on him who the stranger might be. He eased down the deck, gave his hunch to Captain Smith and asked if he should do what Ismay wanted. Smith answered a crisp, 'Carry on.' Returning to No. 5, Pitman jumped in and called, 'Come along, ladies!'

Mrs Catherine Crosby and her daughter Harriet were firmly propelled into the boat by her husband, Captain Edward Gifford Crosby, a Milwaukee shipping man and an old Great Lakes skipper. Captain Crosby had a way of knowing things – right after the crash he scolded his wife, 'You'll lie there and drown!' Later he told her, 'This ship is badly damaged, but I think the watertight compartments will hold her up.' Now he was taking no chances.

Slowly others edged forward – Miss Helen Ostby . . . Mrs F. M. Warren . . . Mrs Washington Dodge and her five-year-old son . . . a young stewardess. When no more women would go alone, a few couples were allowed. Then a few single men. On the starboard side this was the rule all evening – women first, but men if there was still room.

Just aft, First Officer Murdoch, in charge of the starboard side, was having the same trouble filling No. 7. Serial movie star Dorothy Gibson jumped in, followed by her mother. Then they persuaded their bridge companions of the evening, William Sloper and Fred Seward, to join them. Others trickled in, until there were finally nineteen or twenty in the boat. Murdoch felt he could wait no longer. At 12.45 he waved away No. 7 – the first boat down.

Then he ordered Pitman to take charge of No. 5, told him to hang around the after gangway, shook hands and smiled: 'Good-bye, good luck.'

As No. 5 creaked downward, Bruce Ismay was beside himself. 'Lower away! Lower away! Lower away! Lower away!' he chanted, waving one arm in huge circles while hanging on to the davit with the other.

'If you'll get the hell out of the way,' exploded Fifth Officer Lowe, who was working the davits, 'I'll be able to do something! You want me to lower away quickly? You'll have me drown the whole lot of them!'

Ismay was completely abashed. Without a word he turned and walked forward to No. 3.

Old-timers in the crew gasped. They felt Lowe's outburst was the most dramatic thing that could happen tonight. A Fifth Officer doesn't insult the president of the line and get away with it. When they reached New York, there would be a day of reckoning.

And nearly everyone still expected to reach New York. At worst, they would all be transferred to other ships.

'Peuchen,' said Charles M. Hays as the major began helping with the boats, 'this ship is good for eight hours yet. I have just been getting this from one of the best old seamen, Mr Crosby of Milwaukee.'

Monsieur Gatti, *maître* of the ship's *à la carte* French restaurant, was equally unperturbed. Standing alone on the boat deck, he seemed the picture of dignity – his top hat firmly in place, grip in hand and a shawl travelling-blanket folded neatly over his arm.

Mr and Mrs Lucien Smith and Mr and Mrs Sleeper Harper sat quietly chatting in the gym just off the boat deck. The mechanical horses were riderless now – the Astors had moved off somewhere else. And for once there was no one on the stationary bicycles, which the passengers liked to ride, pedalling red and blue arrows round a big white clock. But the room with its bright, blocked linoleum floor and the comfortable wicker chairs was far more pleasant than the boat deck. Certainly it was warmer, and there seemed no hurry.

In the nearly empty smoking-room on A deck, four men sat calmly around a table – Archie Butt, Clarence Moore, Frank Millet, and Arthur Ryerson seemed deliberately trying to avoid the noisy confusion of the boat deck.

Far below, greaser Thomas Ranger began turning off some forty-five electric fans used in the engine room, and he thought about the ones he had to repair tomorrow. Electrician Alfred White, working on the dynamos, brewed some coffee at his post.

At the very stern of the *Titanic*, Quartermaster George Thomas Rowe still paced his lonely watch. He had seen no one, heard nothing since the iceberg glided by nearly an hour ago. Suddenly he was amazed to see a lifeboat floating near the starboard side. He phoned the bridge – did they know there was a boat afloat? An incredulous voice asked who he was. Rowe explained, and the bridge then realized he had been overlooked. They told him to come to the bridge right away and bring some rockets with him. Rowe dropped down to a locker one deck below, picked up a tin box with twelve rockets inside, and clambered forward – the last man to learn what was going on.

Others knew all too well by now. Old Dr O'Loughlin whispered to stewardess Mary Sloan, 'Child, things are very bad.' Stewardess Annie Robinson stood near the mail room, watching the water rise on F deck. As she puzzled over a man's Gladstone bag lying abandoned in the corridor, carpenter Hutchinson arrived with a lead line in his hand – he looked bewildered, distracted, wildly upset. A little later Miss Robinson bumped into Thomas Andrews on A deck. Andrews greeted her like a cross parent:

'I thought I told you to put your lifebelt on!'

'Yes,' she replied, 'but I thought it mean to wear it.'

'Never mind that. Put it on; walk about; let the passengers see you.'

'It looks rather mean.'

'No, put it on . . . Well, if you value your life, put it on.'

Andrews understood people very well. A charming, dynamic man, he was everywhere, helping everyone. And people looked to him. He handled them differently, depending on what he thought of them. He told garrulous steward Johnson that everything would be all right.

He told Mr and Mrs Albert Dick, his casual dinner companions, 'She is torn to bits below, but she will not sink if her after bulkheads hold.' He told competent stewardess Mary Sloan, 'It is very serious, but keep the bad news quiet, for fear of panic.' He told John B. Thayer, whom he trusted implicitly, that he didn't give the ship 'much over an hour to live'.

Some of the crew didn't need to be told. About 12.45, able seaman John Poingdestre left the boat deck to get his rubber boots. He found them in the forecastle on E deck forward, and was just starting up again when the wooden wall between his quarters and some third-class space to starboard suddenly gave way. The sea surged in, and he fought his way out through water up to his waist.

Further aft, dining-saloon steward Ray went to his quarters on E deck to get a warmer overcoat. Coming back up, he went forward on 'Scotland Road' towards the main staircase. The jostling firemen and third-class passengers were gone now. All was quiet along the broad working alleyway, except for water sloshing along the corridor from somewhere forward.

Still further aft, assistant second steward Joseph Thomas Wheat dropped down to pick up some valuables from his room on F deck, port side. It was right next to the Turkish bath, a gloriously garish set of rooms that formed a sort of bridge between the Victorian and Rudolph Valentino eras of interior decoration. The mosaic floor, the blue-green tiled walls, the gilded beams in the dull red ceiling, the stanchions encased in carved teak – all were still perfectly dry.

But when Wheat walked a few yards down the corridor and started back up the stairs, he saw a strange sight: a thin stream of water was flowing *down* the stairs from E deck above. It was only a quarter-inch deep – just about covered the heel of his shoe – as he splashed up the steps. When he reached E deck, he saw it was coming from the starboard side forward.

He guessed what had happened: water forward on F deck, blocked by the watertight compartment door, had risen to E deck, where there was no door, and now was slopping over into the next compartment aft.

Boiler room No. 5 was the only place where everything seemed under control. After the fires were drawn, leading fireman Barrett sent most of the stokers up to their boat stations. He and a few others stayed behind to help engineers Harvey and Shepherd with the pumps.

At Harvey's orders he lifted the iron manhole cover off the floor plates on the starboard side, so Harvey could get at the valves to adjust the pumps.

The boiler room was now clouding up with steam from the water used to wet down the furnaces. In the dim light of their own private Turkish bath, the men worked on . . . vague shapes moving about through the mist.

Then Shepherd, hurrying across the room, fell into the manhole and broke his leg. Harvey, Barrett and fireman George Kemish rushed over. They lifted him up and carried him to the pump room, a closed-off space at one end of the boiler room.

No time to do more than make him comfortable . . . then back into the clouds of steam. Soon orders came down from the bridge for all hands to report to boat stations. As the men went up, Shepherd still lay in the pump room; Barrett and Harvey kept working with the valves. Another fifteen minutes and both men were beginning to cheer up – the room was still dry, the rhythm of the pumps was fast and smooth.

Suddenly the sea came roaring through the space between the boilers at the forward end of the room. The whole bulkhead between No. 5 and No. 6 collapsed.

Harvey shouted to Barrett to make for the escape ladder. Barrett scrambled up, the foam surging around his feet. Harvey himself turned towards the pump room where Shepherd lay. He was still heading there when he disappeared under the torrent of rising water.

The silence in the Marconi shack was broken only by the rasping spark of the wireless, as Phillips rapped out his call for help and took down the answers that bounced back. Bride was still struggling into his clothes, between dashes to the bridge.

So far the news was encouraging. First to reply was the North German Lloyd steamer *Frankfort*. At 12.18 she sent a crisp 'OK . . . Stand by' – but no position. In another minute acknowledgements were pouring in – the Canadian Pacific's *Mt Temple* . . . the Allan liner *Virginian* . . . the Russian tramp *Birma*.

The night crackled with signals. Ships out of direct contact got the word from those within range . . . The news spread in ever-widening circles. Cape Race heard it directly and relayed it inland. On the roof of Wanamaker's department store in New York, a young wireless operator named David Sarnoff caught a faint signal and also passed it on. The whole world was snapping to agonized attention.

Close at hand, the Cunarder *Carpathia* steamed southward in complete ignorance. Her single wireless operator, Harold Thomas Cottam, was on the bridge when Phillips sent his CQD. Now Cottam was back at his set and thought he'd be helpful. Did the *Titanic* know, he casually asked, that there were some private messages waiting for her from Cape Race?

It was 12.25 when Phillips tapped back an answer that brushed aside the *Carpathia*'s courteous gesture: 'Come at once. We have struck a berg. It's a CQD, old man. Position 41.46N 50.14W.'

A moment of appalled silence . . . then Cottam asked whether to tell his captain. Phillips: 'Yes, quick.' Another five minutes and welcome news – the *Carpathia* was only fifty-eight miles away and 'coming hard'.

At 12.34 it was the *Frankfort* again – she was 150 miles away. Phillips asked, 'Are you coming to our assistance?' *Frankfort*: 'What is the matter with you?' Phillips: 'Tell your captain to come to our help. We are on the ice.'

Captain Smith now dropped into the shack for a first-hand picture. The *Olympic*, the *Titanic*'s huge sister ship, was just chiming in. She was 500 miles away; but her set was powerful, she could handle a major rescue job and there was a strong bond between the two liners. Phillips kept in close touch, while urging on the ships that were nearer.

'What call are you sending?' Smith asked.

'CQD,' Phillips answered noncommitally.

Bride had a bright idea. While CQD was the traditional distress call, an international convention had just agreed to use instead the letters SOS – they were easy for the rankest amateur to pick up. So Bride suggested: 'Send SOS; it's the new call, and it may be your last chance to send it.'

Phillips laughed at the joke and switched the call. The clock in the wireless shack said 12.45 a.m. when the *Titanic* sent the first SOS ever flashed by an ocean liner.

None of the ships contacted seemed as promising as the light that winked ten miles off the *Titanic*'s port bow. Through his binoculars Fourth Officer Boxhall saw clearly that it was a steamer. Once, as he tried to get in touch with the Morse lamp, he felt he saw an answer. But he could make nothing of it and finally decided it must be her mast light flickering.

Stronger measures were necessary. As soon as Quartermaster Rowe reached the bridge, Captain Smith asked if he had brought the rockets. Rowe produced them, and the Captain ordered, 'Fire one, and fire one every five or six minutes.'

At 12.45 a blinding flash seared the night. The first rocket shot up from the starboard side of the bridge. Up . . . up it soared, far above the lacework of masts and rigging. Then with a distant, muffled report it burst, and a shower of bright white stars floated down towards the sea. In the blue-white light Fifth Officer Lowe remembered catching a glimpse of Bruce Ismay's startled face.

Ten miles away, apprentice James Gibson stood on the

bridge of the *Californian.* The strange ship that came up from the east had not moved for an hour, and Gibson studied her with interest. With glasses he could make out her side lights and a glare of lights on her afterdeck. At one point he thought she was trying to signal the *Californian* with her Morse lamp. He tried to answer with his own lamp, but soon gave up. He decided the stranger's masthead light was merely flickering.

Second Officer Herbert Stone, pacing the *Californian*'s bridge, also kept his eye on this strange steamer. At 12.45 he saw a sudden flash of white light burst over her. Strange, he thought, that a ship would fire rockets at night.

4. 'You Go and I'll Stay a While'

Second-class passenger Lawrence Beesley considered himself the rankest landlubber, but even he knew what rockets meant. The *Titanic* needed help – needed it so badly she was calling on any ship near enough to see.

The others on the boat deck understood too. There was no more joking or lingering. In fact, there was hardly time to say good-bye.

'It's all right, little girl,' called Dan Marvin to his new bride; 'you go and I'll stay a while.' He blew her a kiss as she entered the boat.

'I'll see you later,' Adolf Dyker smiled as he helped Mrs Dyker across the gunwale.

'Be brave; no matter what happens, be brave,' Dr W. T. Minahan told Mrs Minahan as he stepped back with the other men.

Mr Turrell Cavendish said nothing to Mrs Cavendish. Just a kiss . . . a long look . . . another kiss . . . and he disappeared into the crowd.

Mark Fortune took his wife's valuables, as he and his son Charles saw off Mrs Fortune and their three daughters. 'I'll take care of them; we're going in the next boat,' he explained.

'Charles, take care of Father,' one of the girls called back to her brother.

'Walter, you must come with me,' begged Mrs Walter D. Douglas.

'No,' Mr Douglas replied, turning away, 'I must be a gentleman.'

'Try and get off with Major Butt and Mr Moore,' came a final bit of wifely advice. 'They are big, strong fellows and will surely make it.'

On the fringe of the crowd stood a young Spanish honeymoon couple. Señor Victor de Satode Penasco was just eighteen years old and his bride only seventeen. Neither could understand English. As they watched in bewilderment, the Countess of Rothes spied them and hurried over. A few hurried words in French . . . then Señor Penasco delivered his bride to the countess's care and stepped back into the shadows.

Some of the wives still refused to go. Mr and Mrs Edgar Meyer of New York felt so self-conscious arguing about it in public that they went down to their cabin. There, they decided to part on account of their baby.

Arthur Ryerson had to lay down the law to Mrs Ryerson: 'You must obey orders. When they say "Women and children to the boats," you *must* go when your turn comes. I'll stay here with Jack Thayer. We'll be all right.'

Alexander T. Compton Jr was just as firm when his mother announced she would stay rather than leave him behind: 'Don't be foolish, Mother. You and Sister go in the boat – I'll look out for myself.'

Mr and Mrs Lucien Smith were having the same kind of argument. Seeing Captain Smith standing near with a megaphone, Mrs Smith had an inspiration. She went up to him, explained she was all alone in the world, and asked if her husband could go along with her. The old captain ignored her, lifted his megaphone and shouted, 'Women and children first!'

At this point Mr Smith broke in: 'Never mind, Captain, about that; I'll see she gets in the boat.' Turning to his wife, he spoke very slowly: 'I never expected to ask you to obey, but this is one time you must. It is only a matter of form to have women and children first. The ship is thoroughly equipped and everyone on her will be saved.'

Mrs Smith asked him if he was being completely truthful. Mr Smith gave a firm, decisive 'Yes'. So they kissed good-bye, and as the boat dropped to the sea, he called from the deck, 'Keep your hands in your pockets; it is very cold weather.'

Sometimes it took more than gentle deception. Mrs Emil Taussig was clinging to her husband when No. 8 started down with her daughter. Mrs Taussig turned and cried, 'Ruth!' The brief distraction proved enough: two men tore her from Mr Taussig and dropped her into the lowering boat.

A seaman yanked Mrs Charlotte Collyer by the arm, another by her waist, and they dragged her from her husband Harvey. As she kicked to get free, she heard him call, 'Go, Lottie! For God's sake, be brave and go! I'll get a seat in another boat!'

When Celiney Yasbeck saw she had to go alone, she began yelling and crying to rejoin Mr Yasbeck, but the boat dropped to the sea while she tried in vain to get out.

No amount of persuasion or force could move Mrs

Hudson J. Allison of Montreal. A little apart from the rest, she huddled close to Mr Allison. Their baby Trevor had gone in a boat with the nurse, but Lorraine, their three-year-old daughter, still tugged at her mother's skirt.

Mrs Isidor Straus also refused to go: 'I've always stayed with my husband; so why should I leave him now?'

They had indeed come a long way together: the ashes of the Confederacy . . . the small china business in Philadelphia . . . building Macy's into a national institution . . . Congress . . . and now the happy twilight that crowned successful life – advisory boards, charities, hobbies, travel. This winter they had been to Cap Martin, and the *Titanic*'s maiden voyage had seemed a pleasant way to finish the trip.

Tonight the Strauses came on deck with the others, and at first Mrs Straus seemed uncertain what to do. At one point she handed some small jewellery to her maid Ellen Bird, then took it back again. Later she crossed the boat deck and almost entered No. 8 – then turned around and rejoined Mr Straus. Now her mind was made up: 'We have been living together for many years. Where you go, I go.'

Archibald Gracie, Hugh Woolner, other friends tried in vain to make her go. Then Woolner turned to Mr Straus: 'I'm sure nobody would object to an old gentleman like you getting in . . .'

'I will not go before the other men,' he said, and that was that. Mrs Straus tightened her grasp on his arm, patted it, smiled up at him, and smiled at the group hovering around them. Then they sat down together on a pair of deck chairs.

But most of the women entered the boats – wives

escorted by their husbands, single ladies by the men who had volunteered to look after them. This was the era when gentlemen formally offered their services to 'unprotected ladies' at the start of an Atlantic voyage. Tonight the courtesy came in handy.

Mrs William T. Graham, nineteen-year-old Margaret and her governess Miss Shutes were helped into boat 8 by Howard Case, London manager of Vacuum Oil, and young Washington Augustus Roebling, the steel heir who was striking out on his own as manager of the Mercer Automobile Works in Trenton, New Jersey. As No. 8 dropped to the sea, Mrs Graham watched Case, leaning against the rail, light a cigarette and wave good-bye.

Mrs E. D. Appleton, Mrs R. C. Cornell, Mrs J. Murray Brown and Miss Edith Evans, returning from a family funeral in Britain, were under Colonel Gracie's wing, but somehow in the crowd he lost them, and it wasn't until much later that he found them again.

Perhaps the colonel was distracted by his simultaneous efforts to look after Mrs Churchill Candee, his table companion in the dining-saloon. Mrs Candee was returning from Paris to see her son, who had suffered the novelty of an aeroplane accident, and she must have been attractive indeed. Just about everybody wanted to protect her.

When Edward A. Kent, another table companion, found her after the crash, she gave him an ivory miniature of her mother for safekeeping. Then Hugh Woolner and Bjornstrom Steffanson arrived and helped her into boat 6. Woolner waved good-bye, assuring her that they would help her on board again when the *Titanic* 'steadied herself'. A little later Gracie and Clinch Smith dashed up, also in search of Mrs Candee, but Woolner told them, perhaps a little smugly, that she had been cared for and was safely away.

It was just as well, for the slant in the deck was steeper, and even the carefree were growing uneasy. Some who left everything in their cabins now thought better of it and ventured below to get their valuables. They were in for unpleasant surprises. Celiney Yasbeck found her room was completely under water. Gus Cohen discovered the same thing. Victorine, the Ryersons' French maid, had an even more disturbing experience. She found her cabin still dry, but as she rummaged about, she heard a key turn, and suddenly realized the steward was locking the stateroom door to prevent looting. Her shriek was just in time to keep him from locking her in. Without stretching her luck any further, she dashed back on deck empty-handed.

Time was clearly running out. Thomas Andrews walked from boat to boat, urging the women to hurry: 'Ladies, you *must* get in at once. There is not a moment to lose. You cannot pick and choose your boat. Don't hesitate. Get in, get in!'

Andrews had good reason to be exasperated. Women were never more unpredictable. One girl waiting to climb into No. 8 suddenly cried out, 'I've forgotten Jack's photograph and must get it.' Everybody protested, but she darted below. In a moment she reappeared with the picture and was rushed into the boat.

It was all so urgent – and yet so calm – that Second

Officer Lightoller felt he was wasting time when Chief Officer Wilde asked him to help find the firearms. Quickly he led the captain, Wilde and First Officer Murdoch to the locker where they were kept. Wilde shoved one of the guns into Lightoller's hand, remarking, 'You may need it.' Lightoller stuck it in his pocket and hurried back to the boat.

One after another they now dropped rapidly into the sea: No. 6 at 12.55 ... No. 3 at 1.00 ... No. 8 at 1.10. Watching them go, first-class passenger William Carter advised Harry Widener to try for a boat. Widener shook his head : 'I think I'll stick to the big ship, Billy, and take a chance.'

Some of the crew weren't as optimistic. When assistant second steward Wheat noticed chief steward Latimer wearing his lifebelt over his greatcoat, he urged the chief to put it under the coat – this made swimming easier.

On the bridge, as Fourth Officer Boxhall and Quartermaster Rowe fired off more rockets, Boxhall still couldn't believe what was happening. 'Captain,' he asked, 'is it *really* serious?'

'Mr Andrews tells me,' Smith answered quietly, 'that he gives her from an hour to an hour and a half.'

Lightoller had a more tangible yardstick – the steep, narrow emergency staircase that ran from the boat deck all the way down to E deck. The water was slowly crawling up the stairs, and from time to time Lightoller walked over to the entrance and checked the number of steps it had climbed. He could see very easily, for the lights still gleamed under the pale green water.

His gauge showed time was flying. The pace grew faster – and sloppier. A pretty French girl stumbled and fell as she tried to climb into No. 9. An older woman in a black dress missed No. 10 entirely. She fell between the boat and the side of the ship. But as the crowd gasped, someone miraculously caught her ankle. Others hauled her into the promenade deck below and she climbed back to the boat deck for another try. This time she made it.

Some of them lost their nerve. An old lady made a big fuss at No. 9, finally shook off everybody, and ran away from the boat altogether. A hysterical woman thrashed about helplessly, trying to climb into No. 11. Steward Witter stood on the rail to help her, but she lost her footing anyway, and they tumbled into the boat together. A large fat woman stood crying near No. 13: 'Don't put me in the boat. I don't want to go into the boat! I have never been in an open boat in my life!'

Steward Ray brushed aside her protests – 'You've got to go, and you may as well keep quiet.'

A plan to fill some of the boats from the lower gangways went completely haywire. The doors that were to be used were never opened. The boats that were to stand by rowed off. The people who were to go were left stranded. When the Caldwells and several others went all the way down to a closed gangway on C deck, somebody who didn't know about the plan locked the door behind them. Later some men on the deck above discovered the group and lowered a ladder for them to crawl back up.

A shortage of trained seamen made the confusion worse. Some of the best men had been used to man the early boats. Other old hands were off on special jobs – rounding up lanterns, opening the A deck windows, helping fire the rockets. Six seamen were lost when they went down to open one of the lower gangways; they never came back . . . probably trapped far below. Now Lightoller was rationing the hands he had left – only two crewmen to a lifeboat.

No. 6 was halfway down when a woman called up to the boat deck, 'We've only one seaman in the boat!'

'Any seamen there?' Lightoller asked the people on deck.

'If you like, I will go,' called a voice from the crowd.

'Are you a seaman?'

'I am a yachtsman.'

'If you're sailor enough to get out on that fall, you can go down.' Major Arthur Godfrey Peuchen – Vice-Commodore of the Royal Canadian Yacht Club – swung himself out on the forward fall and slid down into the boat. He was the only male passenger Lightoller allowed in a boat that night.

Men had it luckier on the starboard side. Murdoch continued to allow them in if there was room. The French aviator Pierre Maréchal and sculptor Paul Chevré climbed into No. 7. A couple of Gimbels buyers reached No. 5. When the time came to lower No. 3, Henry Sleeper Harper not only joined his wife, but he brought along his Pekingese Sun Yat-Sen and an Egyptian dragoman named Hamad Hassah, whom he had picked up in Cairo as a sort of joke.

On the same side, Dr Washington Dodge was standing

uncertainly in the shadow of No. 13, when dining-room steward Ray noticed him. Ray asked whether the doctor's wife and son were off, and Dodge said yes. Ray was relieved, because he took a personal interest in them. He had served the Dodges coming over on the *Olympic*. In fact, he was why the Dodges were on board . . . It was no time for philosophy – Ray called out, 'You had better get in here,' and he pushed the doctor into the boat.

The scene was almost punctilious at No. 1. Sir Cosmo Duff Gordon, his wife and her secretary Miss Francatelli – whom Lady Duff Gordon liked to call Miss Franks – asked Murdoch if they could enter.

'Oh, certainly do; I'll be very pleased,' Murdoch replied, according to Sir Cosmo. (On the other hand, lookout George Symons, standing near, thought Murdoch merely said, 'Yes, jump in.') Then two Americans, Abraham Solomon and C. E. H. Stengel, came up and were invited in too. Stengel had trouble climbing over the rail; finally got on top of it and rolled into the boat. Murdoch, an agile terrier of a man, laughed pleasantly, 'That's the funniest thing I've seen tonight.'

Nobody else seemed to be around – all the nearby boats were gone and the crowd had moved aft. When the five passengers were safely loaded, Murdoch added six stokers, put lookout Symons in charge and told him, 'Stand off from the ship's side and return when we call you.' Then he waved to the men at the davits, and they lowered No. 1 – capacity forty persons – with exactly twelve people.

As the boat creaked down, greaser Walter Hurst

watched it from the forward well deck. He remembers observing somcwhat caustically, 'If they are sending the boats away, they might just as well put some people in them.'

Down in third class there were those who didn't even have the opportunity to miss going in No. 1. A swarm of men and women milled around the foot of the main steerage staircase, all the way aft on E deck. They had been there ever since the stewards got them up. At first there were just women and married couples; but then the men arrived from forward, pouring back along 'Scotland Road' with their luggage. Now they were all jammed together – noisy and restless, looking more like inmates than passengers amid the low ceilings, the naked light bulbs, thc scrubbed simplicity of the plain white walls.

Third-class steward John Edward Hart struggled to get them into life jackets. He didn't have much luck – partly because he was also assuring them there was no danger, partly because many of them didn't understand English anyhow. Interpreter Muller did the best he could with the scores of Finns and Swedes, but it was slow going.

At 12.30 orders came down to send the women and children up to the boat deck. It was hopeless to expect them to find their way alone through the maze of passages normally sealed off from third class; so Hart decided to escort them up in little groups. This took time too, but at last a convoy was organized and started off.

It was a long trip – up the broad stairs to the third-class lounge on C deck . . . across the open well deck . . . by the second-class library and into first-class quarters. Then down the long corridor by the surgeon's office, the private saloon for the maids and valets of first-class passengers, finally up the grand stairway to the boat deck.

Hart led his group to boat No. 8, but even then the job wasn't over. As fast as he got them in, they would jump out and go inside where it was warm.

It was after one o'clock when Hart got back to E deck to organizc another trip. It was no easier. Many women still refused to go. On the other hand, some of the men now insisted on going. But that was out of the question, according to the orders he had.

Finally he was off again on the same long trek. It was 1.20 by the time he reached the boat deck and led the group to No. 15. No time to go back for more. Murdoch ordered him into the boat and off he went with his second batch at about 1.30.

There was no hard-and-fast policy. One way or another, many of the steerage passengers avoided the cul-de-sac on E deck and got topside. There they stood waiting, nobody to guide or help them. A few of the barriers that marked off their quarters were down. Those who came across these openings wandered into other parts of the ship. Some eventually found their way to the boat deck.

But most of the barriers were not down, and the steerage passengers who sensed danger and aimed for the boats were strictly on their own resources.

Like a stream of ants, a thin line of them curled their way up a crane in the after well deck, crawled along the boom to the first-class quarters, then over the railing and on up to the boat deck.

Some slipped under a rope that had been stretched across the after well deck, penning them even further to the stern than the regular barrier. But once through, it was fairly easy to get to the second-class stairway and on up to the boats.

Others somehow reached the second-class promenade space on B deck, then couldn't find their way any further. In desperation they turned to an emergency ladder meant for the crew's use. This ladder was near the brightly lit windows of the first-class *à la carte* restaurant, and as Anna Sjoblom prepared to climb up with another girl, they looked in. They marvelled at the tables beautifully set with silver and china for the following day. The other girl had an impulse to kick the window out and go inside, but Anna persuaded her that the company might make them pay for the damage.

Many of the steerage men climbed another emergency ladder from the forward well deck, and then up the regular first-class companionway to the boats.

Others beat on the barriers, demanding to be let through. As third-class passenger Daniel Buckley climbed some steps leading to a gate to first class, the man ahead of him was chucked down by a seaman standing guard. Furious, the passenger jumped to his feet and raced up the steps again. The seaman took one look, locked the gate and fled. The passenger smashed the lock and dashed through, howling what he would do if he caught the sailor. With the gate down, Buckley and dozens of others swarmed into first class. At another barrier a seaman held back Kathy Gilnagh, Kate Mullins and Kate Murphy. (On the *Titanic* all Irish girls seemed to be named Katherine.) Suddenly steerage passenger Jim Farrell, a strapping Irishman from the girls' home country, barged up. 'Great God, man!' he roared. 'Open the gate and let the girls through!' It was a superb demonstration of sheer voice-power. To the girls' astonishment the sailor meekly complied.

Even then, Kathy Gilnagh's troubles weren't over. She took a wrong turn . . . lost her friends . . . found herself alone on the second-class promenade, with no idea how to reach the boats. The deck was deserted, except for a single man leaning against the rail, staring moodily into the night. He let her stand on his shoulders, and she managed to climb to the next deck up. When she finally reached the boat deck, No. 16 was just starting down. A man warned her off – there was no more room.

'But I want to go with my sister!' Kathy cried. She had no sister, but it seemed a good way to move the man. And it worked. 'All right, get in,' he sighed, and she slipped into the boat as it dropped to the sea – another third-class passenger safely away.

But for every steerage passenger who found an escape, hundreds milled aimlessly around the forward well deck . . . the after poop deck . . . or the foot of the E deck staircase. Some stayed in their cabins – that's where Mary Agatha Glynn and four discouraged room-mates were found by young Martin Gallagher. He quickly escorted them to boat 13 and stepped back on the deck again. Others turned to prayer. When steerage passenger Gus Cohen passed the third-class dining-saloon about an hour after the crash, he saw quite a number gathered there, many with rosaries in their hands.

The staff of the first-class *à la carte* restaurant were having the hardest time of all. They were neither fish nor fowl. Obviously they weren't passengers, but technically they weren't crew either. The restaurant was not run by the White Star but by Monsieur Gatti as a concession.

Thus, the employees had no status at all. And to make matters worse, they were French and Italian – objects of deep Anglo-Saxon suspicion at a time like this in 1912.

From the very start they never had a chance. Steward Johnson remembered seeing them herded together down by their quarters on E deck aft. Manager Gatti, his chef and the chef's assistant, Paul Maugé, were the only ones who made it to the boat deck. They got through because they happened to be in civilian clothes; the crew thought they were passengers.

Down in the engine room no one even thought of getting away. Men struggled desperately to keep the steam up . . . the lights lit . . . the pumps going. Chief engineer Bell had all the watertight doors raised aft of boiler room No. 4 – when the water reached here they could be lowered again; meanwhile it would be easier to move around.

Greaser Fred Scott worked to free a shipmate trapped in the after tunnel behind one of the doors. Greaser Thomas Ranger turned off the last of the forty-five ventilating fans – they used too much electricity. Trimmer Thomas Patrick Dillon helped to drag long sections of pipe from the aft compartments, to get more volume out of the suction pump in boiler room No. 4.

Here, trimmer George Cavell was busy drawing the fires. This meant even less power, but there must be no explosion when the sea reached No. 4. It was about 1.20 and the job was almost done when he noticed the water seeping up through the metal floor plates. Cavell worked faster. When it reached his knees, he had had enough. He was almost at the top of the escape ladder when he began to feel he had quit on his mates. Down again, only to find they were gone too. His conscience clear, he climbed back up, this time for good.

Most of the boats were now gone. One by one they rowed slowly away from the *Titanic*, oars bumping and splashing in the glass-smooth sea.

'I never had an oar in my hand before, but I think I can row,' a steward told Mrs J. Stuart White, as No. 8 set out.

In every boat all eyes were glued on the *Titanic*. Her tall masts, the four big funnels stood out sharp and black in the clear blue night. The bright promenade decks, the long rows of portholes all blazed with light. From the boats they could see the people lining the rails; they could hear the ragtime in the still night air. It seemed impossible that anything could be wrong with this great ship; yet there they were out on the sea, and there she was, well down at the head. Brilliantly lit from stem to stern, she looked like a sagging birthday cake.

Clumsily the boats moved further away. Those told to stand by now lay on their oars. Others, told to make for the steamer whose lights shone in the distance, began their painful journey.

The steamer seemed agonizingly near. So near that

Captain Smith told the people in boat 8 to go over, land its passengers, and come back for more. About the same time he asked Quartermaster Rowe at the rocket gun if he could Morse. Rowe replied he could a little, and the captain said, 'Call that ship up and when she replies, tell her, "We are the *Titanic* sinking; please have all your boats ready."'

Boxhall had already tried to reach her, but Rowe was more than eager to try his own luck; so in between rocket firing he called her again and again. Still no answer. Then Rowe told Captain Smith he thought he saw another light on the starboard quarter. The old skipper squinted through his glasses, courteously told Rowe that it was a planet. But he liked the eagerness of his young quartermaster, and he lent Rowe the glasses to see for himself.

Meanwhile Boxhall continued firing rockets. Sooner or later, somehow they would wake up the stranger.

On the bridge of the *Californian*, Second Officer Stone and apprentice Gibson counted the rockets – five by 12.55. Gibson tried his Morse lamp again, and at one o'clock lifted his glasses for another look. He was just in time to see a sixth rocket.

At 1.10 Stone whistled down the speaking tube to the chart room and told Captain Lord. He called back, 'Are they company signals?'

'I don't know,' Stone answered, 'but they appear to me to be white rockets.'

The captain advised him to go on Morsing.

A little later Stone handed his glasses to Gibson,

remarking: 'Have a look at her now. She looks very queer out of the water – her lights look queer.'

Gibson studied the ship carefully. She seemed to be listing. She had, as he called it, 'a big side out of the water'. And Stone, standing beside him, noticed that her red side light had disappeared.

1. The *Titanic* on the stocks at Harland & Woolf, Belfast
(Harland & Woolf)

2. The *Titanic* outward bound, 10 April 1912
(Beken of Cowes)

3. The *Titanic* at Cherbourg on the evening of 10 April. Although heavily retouched, this view suggests how she looked to those coming aboard by tender (author)

4. When the crash came, most of the first-class passengers still up were in the smoking-room. It was apparently never photographed, but this view taken on the *Titanic*'s sister ship *Olympic* suggests the elegant setting (The Byron Collection, City of New York Museum)

5. The Café Parisien combined sturdy British wicker with French *joie de vivre*. It was a favourite with the ship's younger set, and this night was no exception (Harland & Woolf)

6. Special cabin B-59, furnished in Dutch style. For many passengers in staterooms like this, a steward's polite knock on the door was the first hint of trouble (Harland & Woolf)

7. The boat deck, looking forward. The boats on the right are Nos. 9, 11 and 13, among the last lowered (author)

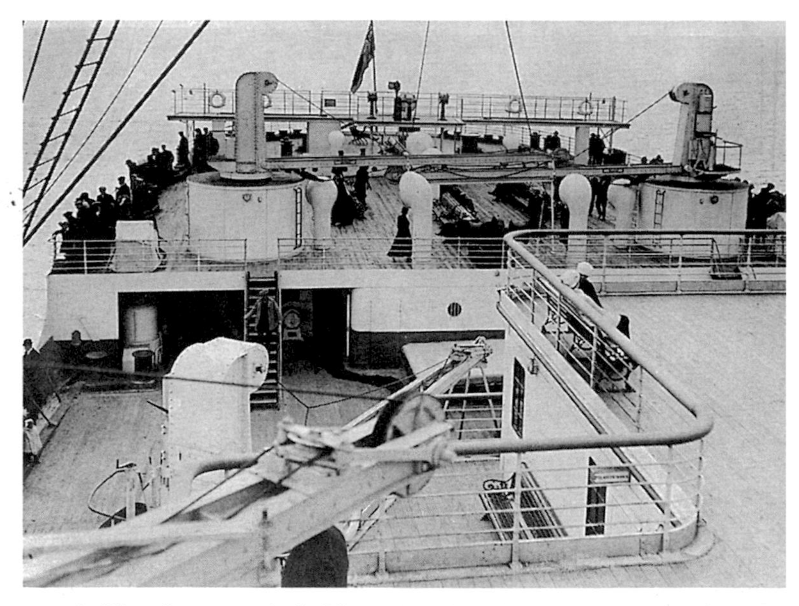

8. The after poop deck. Normally third-class space, more and more people crowded here as the bow sank lower. The rail at the extreme left is where Baker Joughin climbed out on to the side of the ship as she nosed down for the final plunge (Brown Brothers)

9. *Titanic* passengers watch the Queenstown tender come alongside, photographed on 11 April. Three days later the Countess of Rothes handled the tiller of the lifeboat nearest the camera (Underwood & Underwood)

10. As the *Titanic* sideswiped the iceberg, topside it looked like a close shave. Far below, they knew better – it cut a 300-foot gash (*Illustrated London News*)

11. About 1.40 a.m. Last rocket going up ... well deck almost awash ... forecastle head close to the water. Sketched later by steward Leo James Hyland (author)

12. The lights blinked and went out for ever. The forward funnel fell, washing collapsible B clear of the ship (*Harper's Weekly*)

13. (*overleaf*) 'Hanging vertical' (*Illustrated London News*)

5. 'I Believe She's Gone, Hardy'

The other ships just didn't seem to understand. At 1.25 the *Olympic* asked, 'Are you steering south to meet us?' Phillips patiently explained, 'We are putting the women off in the boats.'

Then the *Frankfort*: 'Are there any ships around you already?' Phillips ignored this one. Again the *Frankfort*, asking for more details. This was too much. He jumped up, almost screaming: 'The damn fool! He says, "What's up, old man?"' Then he angrily tapped back: 'You fool, stand by and keep out.'

From time to time Captain Smith dropped in – once to warn that the power was fading . . . again to say she couldn't last much longer . . . later to report that the water had reached the engine room. At 1.45, Phillips begged the *Carpathia*: 'Come as quickly as possible, old man; engine room filling up to the boilers.'

Meanwhile Bride draped an overcoat over Phillips' shoulders, then managed to strap a lifebelt on him. The problem of getting him into his boots was more complicated. Phillips asked whether any boats were left – maybe the boots wouldn't be needed.

Once he turned the set over to Bride, went out to see what was happening. He returned shaking his head: 'Things look very queer.'

They looked queer indeed. The sea now slopped over the *Titanic*'s forward well deck . . . rippled around the cranes, the hatches, the foot of the mast . . . washed against the base of the white superstructure. The roar of steam had died, the nerve-racking rockets had stopped – but the slant of the deck was steeper and there was an ugly list to port.

About 1.40, Chief Officer Wilde shouted, 'Everyone on the starboard side to straighten her up!' Passengers and crew trooped over, and the *Titanic* swung sluggishly back on even keel. The work on the boats resumed.

As No. 2 prepared to cast off at 1.45, steward Johnson, his pockets bulging with oranges, yelled up to the boat deck for a razor to cut the falls. Seaman McAuliffe dropped one down, calling, 'Remember me at Southampton and give it back to me!' McAuliffe was probably the last man on the *Titanic* so sure of returning to Southampton.

First Officer Murdoch knew better. As he walked along the deck with chief steward Hardy of second class, he sighed, 'I believe she's gone, Hardy.'

There was no longer any difficulty persuading people to leave the ship. Paul Maugé, the chef's assistant, jumped ten feet into a dangling boat. Somebody on a lower deck tried to drag him out, but he squirmed free and was safe.

Third-class passenger Daniel Buckley – safely through the broken gate and on to the boat deck – took no more chances. With several other men he jumped into a boat and huddled there crying. Most of the men were hauled out, but from somewhere he got a woman's shawl. He said Mrs Astor put it over him. In any case, the disguise worked.

Another young man – no more than a boy – wasn't as lucky. Fifth Officer Lowe caught him under a seat in No. 14 begging that he wouldn't take up much room. Lowe drew his gun, but the boy only pleaded harder. Then Lowe changed tactics, told him to be a man, and somehow got him out. By now Mrs Charlotte Collyer and other women in the boat were sobbing, and her eight-year-old daughter Marjory joined the uproar, tugging at Lowe's arm and crying, 'Oh, Mr Man, don't shoot, please don't shoot the poor man!'

Lowe paused long enough to smile and nod at her reassuringly. The boy was out now, anyhow, lying face down near a coil of rope.

But No. 14's troubles weren't over. Another wave of men rushed the boat. Seaman Scarrott beat them back with the tiller. This time Lowe pulled his gun and shouted, 'If anyone else tries that, this is what he'll get!' He fired three times along the side of the ship as the boat dropped down to the sea.

Murdoch barely stopped a rush at No. 15. He yelled at the crowd, 'Stand back! Stand back! It's women first!'

All the way forward, there was more trouble at collapsible C, which had been fitted into the davits used by No. 1. A big mob pushed and shoved, trying to climb aboard.

Two men dropped in. Purser Herbert McElroy fired twice into the air. Murdoch shouted, 'Get out of this! Clear out of this!' Hugh Woolner and Bjornstrom Steffanson – attracted by the pistol flashes – rushed over to help. Yanking the culprits by arms, legs, anything, they cleared the boat. The loading continued.

Jack Thayer stood off to one side with Milton Long, a young shipboard acquaintance from Springfield, Massachusetts. They had met for the first time this evening over after-dinner coffee. Following the crash, Long – who was travelling alone – attached himself to the Thayer family, but he and Jack lost the older Thayers in the crowd on A deck. Now they were alone, debating what to do, supposing the rest of the family was already off in the boats. They finally decided to stay clear of boat C. With all the uproar, it seemed bound to tip over.

But they were wrong. Things gradually straightened out, and finally boat C was ready for lowering. Chief Officer Wilde shouted to know who was in command. Hearing him, Captain Smith turned to Quartermaster Rowe – still fiddling with the Morse lamp – and told him to take charge. Rowe jumped in and got ready to lower.

Close by, President Bruce Ismay stood, helping to get the boat ready for lowering. He was calmer now than in those early moments when Lowe had bawled him out – in fact he seemed every inch an accepted member of the *Titanic*'s crew.

This was a frequent role for Ismay, but by no means his only one. Sometimes he preferred the role of passenger. So far during the voyage he had shifted back and forth several times.

At Queenstown he was a sort of super-captain. He told chief engineer Bell the speed he wanted for various stages of the voyage. He also set the New York arrival time at Wednesday morning, instead of Tuesday night. He didn't consult Captain Smith on this.

Later, at sea, Ismay was mostly a passenger, enjoying the fine cuisine of the *à la carte* restaurant . . . shuffleboard . . . bridge . . . tea and scones in his deck chair on the port side of A deck.

This Sunday he was enough of a member of the crew to see the ice message that arrived from another ship. In the bright, sunny Palm Court – just as the bugler sounded lunch – Captain Smith gave him a warning from the *Baltic*. During the afternoon Ismay (who liked to remind people who he was) fished it out of his pocket and waved it at Mrs Ryerson and Mrs Thayer. In the smoking-room before dinner, while the twilight still glowed through the amber-stained windows, Captain Smith sought and got the message back. Then Ismay walked down to the restaurant, immaculate in his dinner jacket, very much a first-class passenger.

After the crash he went back to being in the crew – up with the captain on the bridge . . . consulting with chief engineer Bell . . . and now, despite the tongue-lashing from Fifth Officer Lowe, shouted orders about the boats.

Then came another switch. At the very last moment he suddenly climbed into boat C. Down it dropped, with forty-two people including Bruce Ismay – just another passenger.

Most of the passengers were different. William T. Stead, independent as ever, sat reading alone in the first-class smoking-room. To fireman Kemish, passing by, he looked as though he planned to stay there whatever happened.

Reverend Robert J. Bateman of Jacksonville stood outside, watching his sister-in-law Mrs Ada Balls enter a boat.

'If I don't meet you again in this world,' he called, 'I will in the next.' Then as the boat jerked down, he took off his necktie and tossed it to her as a keepsake.

George Widener and John B. Thayer leaned against the boat deck rail, quietly talking things over. Contrary to young Jack's guess, his father wasn't safe in a boat and, in fact, didn't have any idea of entering one. A little way off, Archie Butt, Clarence Moore, Arthur Ryerson and Walter Douglas stood silently together. Major Butt was very quiet, had no pistol, took no active part, despite the stories later told that he practically took charge.

Further aft, Jay Yates – described as a gambler hoping to make a maiden-voyage killing – stood alone and friendless. To a woman entering a boat, he handed a page torn from his appointment book. Signed with one of his aliases, the note read, 'If saved, inform my sister Mrs F. J. Adams of Findlay, Ohio. Lost. J. H. Rogers.'

Benjamin Guggenheim had a more detailed message: 'If anything should happen to me, tell my wife I've done my best in doing my duty.'

Actually Guggenheim almost outdid himself. Gone was the sweater that steward Etches made him wear. Also his lifebelt. Instead he and his secretary now stood resplendent in evening clothes. 'We've dressed in our best,' he explained, 'and are prepared to go down like gentlemen.'

There were a few couples too. The Allisons stood smiling on the promenade deck, Mrs Allison grasping little Lorraine with one hand, her husband with the other. The Strauses leaned against the boat deck rail, their arms about each other's waist. A young Western couple waited nearby; when Lightoller asked the girl if he could put her in a boat, she told him cheerfully, 'Not on your life. We started together and, if need be, we'll finish together.'

Archibald Gracie, Clinch Smith, a dozen other first-class men worked with the crew, loading the last boats. As they helped Miss Constance Willard of Duluth, Minnesota, they smiled and told her to be brave. She noticed great beads of sweat stood out on their foreheads.

Lightoller was sweating too. He peeled off his great-coat. Even in sweater and pyjamas, he was wringing wet from hard work. He looked so odd on this bitter-cold night that assistant surgeon Simpson, always a wag, called out, 'Hello, Lights, are you warm?'

The assistant surgeon was with old Dr O'Loughlin, Purser McElroy and Assistant Purser Barker. Lightoller joined them for a moment. They all shook hands and said, 'Good-bye.'

No time for more. A glance down the emergency stairway told Lightoller the water was now on C deck . . . rising fast. But the lights were still bright . . . the music still ragtime . . . the beat still lively.

Only two more boats. One of them, No. 4, had been a headache all night. Over an hour ago Lightoller lowered it to A deck, planning to fill it from there, but the windows were all closed. Then someone noticed the *Titanic*'s sounding spar stuck out directly below the boat. Seaman Sam Parks and storekeeper Jack Foley went down to chop it away, but they had trouble finding an axe. Time was wasting. Lightoller hurried on to the other boats – he'd load this one later.

Meanwhile the passengers waiting to go in No. 4 cooled their heels. And they were very prominent heels. The Astors, Wideners, Thayers, Carters and Ryersons were sticking pretty much together. When Lightoller first ordered the boat loaded, wives, children, maids and nurses went down to the promenade deck to get in as a group. When they found they couldn't, they just stayed put.

Eventually most of the husbands turned up, and for over an hour the cream of New York and Philadelphia society just waited around while the windows were opened and the sounding spar chopped away. Once they were ordered back up to the boat deck, but then second steward Dodd sent them right down again. Exasperated, Mrs Thayer exclaimed, 'Just tell us where to go and we will follow! You ordered us up here and now you're sending us back!'

It was 1.45 when Lightoller returned. Now he stood – one foot in No. 4, the other on an open window sill. Somebody put deck chairs against the rail to serve as steps. The men stood by to pass the women and children through the windows.

John Jacob Astor helped Mrs Astor across the frame, then asked if he could join her. She was, as he put it, 'in a delicate condition'.

'No, sir,' Lightoller replied. 'No men are allowed in these boats until the women are loaded first.'

Astor asked which boat it was, and Lightoller said,

'Number 4.' Colonel Gracie was sure Astor merely wanted to locate his wife later. Lightoller was sure he planned to make a complaint.

Then came the Ryersons' turn. Arthur Ryerson noticed their French maid Victorine had no lifebelt. Quickly he stripped off his own and buckled it on her. When Mrs Ryerson led her son Jack to the window, Lightoller called out, 'That boy can't go!'

Mr Ryerson indignantly stepped forward: 'Of course that boy goes with his mother – he is only thirteen.' So they let him pass, Lightoller grumbling, 'No more boys.'

At 1.55, No. 4 dropped to the sea – just 15 feet below. Mrs Ryerson was shocked to see how far the ship had sunk. She watched the water pour in the big square ports on C deck, sweep around the period furniture of the de luxe suites. Then she looked up at the promenade deck. Mr Ryerson was still standing by the rail with Mr Widener, looking down at the boat. They seemed very quiet.

Only one boat was left. Collapsible D had now been fitted into the davits used by No. 2 and was ready for loading. There was no time to spare. The lights were beginning to glow red. Chinaware was breaking somewhere below. Jack Thayer saw a man lurch by with a full bottle of Gordon's gin. He put it to his mouth and drained it. 'If I ever get out of this,' Thayer said to himself, 'there is one man I'll never see again.' (Actually, he was one of the first survivors Thayer met.)

Lightoller took no chances. Most of the passengers had moved aft, but still – one boat . . . forty-seven seats . . .

1,600 people. He had the crew lock arms in a wide ring around boat D. Only the women could come through.

Two baby boys were brought by their father to the edge of the ring, handed through and placed in the boat. The father stepped back into the crowd. He called himself 'Mr Hoffman' and told people he was taking the boys to visit relatives in America. His name really was Navatril and he was kidnapping the children from his estranged wife.

Henry B. Harris, the theatrical producer, escorted Mrs Harris to the ring, was told he couldn't go any further. He sighed: 'Yes, I know. I will stay.'

Colonel Gracie rushed up with Mrs John Murray Brown and Miss Edith Evans, two of the five 'unprotected ladies' to whom he had offered his services on the trip. He was stopped by the line but saw the women through. They reached boat D just as it was starting down the falls. Miss Evans turned to Mrs Brown: 'You go first. You have children waiting at home.'

Quickly she helped Mrs Brown over the rail. Then someone yelled to lower away, and at 2.05 collapsible D – the last boat of all – started down towards the sea – without Edith Evans.

Directly below, Hugh Woolner and Bjornstrom Steffanson were standing alone by the rail. It had been a hard night – helping Mrs Candee . . . trying to save the Strauses . . . dragging those cowards out of boat C. Now they were on A deck trying to find someone else to help, but the deck was absolutely deserted. The lights had a reddish glow.

'This is getting rather a tight corner,' Woolner remarked,

'let's go through the door at the end.' They walked forward to the open end of the promenade deck. As they came out, the sea poured on to the deck, over their evening pumps and up to their knees. They hopped on to the railing. Nine feet away they saw boat D sliding down the side of the ship. It was now or never.

'Let's make a jump for it!' cried Woolner. 'There's plenty of room in her bow!' Steffanson hurled himself out at the boat, landing head over heels up front. The next second Woolner followed, falling half in, half out. In another instant collapsible D hit the water and cast off. As it pulled away, seaman William Lucas called up to Miss Evans still standing on deck. 'There's another boat going to be put down for you.'

6. 'That's the Way of It at This Kind of Time'

With the boats all gone, a curious calm came over the *Titanic*. The excitement and confusion were over, and the hundreds left behind stood quietly on the upper decks. They seemed to cluster inboard, trying to keep as far away from the rail as possible.

Jack Thayer stayed with Milton Long on the starboard side of the boat deck. They studied an empty davit, using it as a yardstick against the sky to gauge how fast she was sinking. They watched the hopeless efforts to clear two collapsibles lashed to the roof of the officers' quarters. They exchanged messages for each other's families. Sometimes they were just silent.

Thayer thought of all the good times he had had and of all the future pleasures he would never enjoy. He thought of his father and his mother, of his sisters and brother. He felt far away, as though he were looking on from some distant place. He felt very, very sorry for himself.

Colonel Gracie, standing a little way off, felt curiously breathless. Later he rather stuffily explained it was the feeling when '*vox faucibus haesit*, as frequently happened to the old Trojan hero of our schooldays'. At the time he merely said to himself, 'Good-bye to all at home.'

In the wireless shack there was no time for either

self-pity or *vox faucibus haesit*. Phillips was still working the set, but the power was very low. Bride stood by, watching people rummage the officers' quarters and the gym, looking for extra lifebelts.

It was 2.05 when Captain Smith entered the shack for the last time: 'Men, you have done your full duty. You can do no more. Abandon your cabin. Now it's every man for himself.'

Phillips looked up for a second, then bent over the set once more. Captain Smith tried again: 'You look out for yourselves. I release you.' A pause, then he added softly, 'That's the way of it at this kind of time . . .'

Phillips went on working. Bride began to gather up their papers. Captain Smith returned to the boat deck, walked about speaking informally to men here and there. To fireman James McGann: 'Well, boys, it's every man for himself.' Again, to oiler Alfred White: 'Well, boys, I guess it's every man for himself.' To steward Edward Brown: 'Well, boys, do your best for the women and children, and look out for yourselves.' To the men on the roof of the officers' quarters: 'You've done your duty, boys. Now, every man for himself.' Then he walked back on the bridge.

Some of the men took the captain at his word and jumped overboard. Night baker Walter Belford leaped as far out as he could, cannon-balled into the water in a sitting position. He still shudders and sucks his breath sharply when he thinks of the stabbing cold. Greaser Fred Scott, just up from boiler room 4, tried to slide down an empty fall, missed and took a belly-flopper into the sea. He was picked up by boat 4, still standing by the ship but trying to row clear of the barrels and deck chairs that were now hurtling down. Steward Cunningham made a hefty leap and also managed to reach No. 4.

But most of the crew stuck to the ship. On top of the officers' quarters, Lightoller noticed trimmer Hemming at work on one of the tangled collapsibles . . . yet Hemming should have gone long ago as part of the crew in No. 6.

'Why haven't you gone, Hemming?'

'Oh, plenty of time yet, sir.'

Not far away two young stewards idly watched Lightoller, Hemming and the others at work. In the fading light of the boat deck, their starched white jackets stood out as they leaned against the rail, debating how long the ship could last. Scattered around the boat deck, some fifteen first-class bellboys were equally at ease – they seemed pleased that nobody cared any longer whether they smoked. Nearby, gymnasium instructor T. W. McCawley, a spry little man in white flannels, explained why he wouldn't wear a life jacket – it kept you afloat but it slowed you down; he felt he could swim clear more quickly without it.

By the forward entrance to the grand staircase, between the first and second funnel, the band – now wearing life jackets on top of their overcoats – scraped lustily away at ragtime.

The passengers were just as calm, although they too had their jumpers. Frederick Hoyt saw his wife into collapsible D, leaped and swam to where he thought the boat might pass. He guessed well. In a few minutes boat D splashed by and hauled him in. For the rest of the night he sat soaked to the skin, rowing hard to keep from freezing.

But for the most part the passengers merely stood waiting or quietly paced the boat deck. New York and Philadelphia society continued to stick together – John B. Thayer, George and Harry Widener, Duane Williams formed a little knot . . . lesser luminaries like Clinch Smith and Colonel Gracie hovering nearby. Astor remained pretty much alone, and the Strauses sat down on deck chairs.

Jack Thayer and Milton Long debated whether to jump. The davit they were using as a gauge showed the *Titanic* was going much faster now. Thayer wanted to jump out, catch an empty lifeboat fall, slide down and swim out to the boats he could dimly see 500 or 600 yards away. He was a good swimmer. Long, not nearly as good, argued against it and persuaded Thayer not to try.

Further forward, Colonel Gracie lent his penknife to the men struggling with the collapsibles lashed to the officers' quarters. They were having a hard time, and Gracie wondered why.

Some of the third-class passengers had now worked their way up to the boat deck, and others were drifting towards the gradually rising stern. The after poop deck, normally third-class space anyhow, was suddenly becoming attractive to all kinds of people.

Olaus Abelseth was one of those who reached the boat deck. Most of the evening he had been all the way aft with his cousin, his brother-in-law and the two Norwegian girls. With other steerage men and women, they aimlessly waited for someone to tell them what to do.

Around 1.30 an officer opened the gate to first class and ordered the women to the boat deck. At 2.00 the men were allowed up too. Many now preferred to stay where they were – this would clearly be the last point above water. But Abelseth, his cousin and brother-in-law went up on the chance there was still a boat left. The last one was pulling away.

So they just stood there, as worried about being in first class as by the circumstances that had brought them there. Abelseth watched the crew trying to free the collapsibles. Once an officer, searching for extra hands, called, 'Are there any sailors here?'

Abelseth had spent sixteen of his twenty-seven years on the sea and felt he should speak up. But his cousin and brother-in-law pleaded, 'No, let us just stay here together.'

So they did. They felt rather awkward and said very little. It was even more awkward when Mr and Mrs Straus drew near. 'Please,' the old gentleman was saying, 'get into a lifeboat and be saved.'

'No, let me stay with you,' she replied. Abelseth turned and looked the other way.

Within the ship the heavy silence of the deserted rooms had a drama of its own. The crystal chandeliers of the *à la carte* restaurant hung at a crazy angle, but they still burned brightly, lighting the fawn panels of French walnut and the rose-coloured carpet. A few of the little table lights with their pink silk shades had fallen over, and someone was rummaging in the pantry, perhaps for something to fortify himself.

The Louis Quinze lounge with its big fireplace was silent and empty. The Palm Court was equally deserted – one passer-by found it hard to believe that just four hours ago it was filled with exquisitely dressed ladies and gentlemen, sipping after-dinner coffee, listening to chamber music by the same men who now played gay songs on the boat deck above.

The smoking-room was not completely empty. When a steward looked in at 2.10, he was surprised to see Thomas Andrews standing all alone in the room. Andrews' lifebelt lay carelessly across the green cloth top of a card table. His arms were folded over his chest; his look was stunned; all his drive and energy were gone. A moment of awed silence, and the steward timidly broke in: 'Aren't you going to have a try for it, Mr Andrews?'

There was no answer, not even a trace that he heard. The builder of the *Titanic* merely stared aft.

Outside on the decks, the crowd still waited; the band still played. A few prayed with the Reverend Thomas R. Byles, a second-class passenger. Others seemed lost in thought.

There was much to think about. For Captain Smith there were the four ice messages he had received during the day – a fifth, which he may not have seen, told exactly where to expect the berg. And there was the thermometer that fell from forty-three degrees at seven o'clock to thirty-two degrees at ten o'clock. And the temperature of the sea, which dropped to thirty-one degrees at 10.30 p.m.

Wireless operator Jack Phillips could ponder over the sixth ice warning – when the *Californian* broke in at 11.00 p.m. and Phillips told her to shut up. That one never even reached the bridge.

George Q. Clifford of Boston had the rueful satisfaction of remembering that he took out 50,000 dollars' extra life insurance just before the trip.

For Isidor Straus there was the irony of his will. A special paragraph urged Mrs Straus to 'be a little selfish; don't always think only of others.' Through the years she had been so self-sacrificing that he especially wanted her to enjoy life after he was gone. Now the very qualities he admired so much meant he could never have his wish.

Little things too could return to haunt a person at a time like this. Edith Evans remembered a fortune-teller who once told her to 'beware of the water'. William T. Stead was nagged by a dream about somebody throwing cats out of a top-storey window. Charles Hays had prophesied just a few hours earlier that the time would soon come for 'the greatest and most appalling of all disasters at sea'.

Two men perhaps wondered why they were there at all. Archie Butt hadn't wanted to go abroad, but he needed a rest; and Frank Millet had badgered President Taft into sending Butt with a message to the Pope – official business but spring in Rome, too. Chief Officer Wilde didn't plan to be on board either. He was regularly on the *Olympic*, but the White Star Line transferred him at the last minute for this one voyage. They thought his experience would be useful in breaking in the new ship. Wilde had considered it a lucky development.

In the wireless shack Phillips struggled to keep the set going. At 2.10 he sounded two V's – heard faintly by the *Virginian* – as he tried to adjust the spark for better results. Bride made a last inspection tour. He returned to find that a fainting lady had been carried into the shack. Bride got her a chair and a glass of water, and she sat gasping while her husband fanned her. She came to, and the man took her away.

Bride went behind the curtain where he and Phillips slept. He gathered up all the loose money, took a last look at his rumpled bunk, pushed through the curtain again. Phillips still sat hunched over the set, completely absorbed. But a stoker was now in the room, gently unfastening Phillips' life jacket.

Bride leaped at the stoker, Phillips jumped up and the three men wrestled around the shack. Finally Bride wrapped his arms about the stoker's waist, and Phillips swung again and again until the man slumped unconscious in Bride's arms.

A minute later they heard the sea gurgling up the A deck companionway and washing over the bridge. Phillips cried, 'Come on, let's clear out!' Bride dropped the stoker, and the two men ran out on to the boat deck. The stoker lay still where he fell.

Phillips disappeared aft. Bride walked forward and joined the men on the roof of the officers' quarters who were trying to free collapsibles A and B. It was a ridiculous place to stow boats – especially when there were only twenty for 2,207 people. With the deck slanting like this, it had been hard enough launching C and D, the two collapsibles stowed right beside the forward davits. It was impossible to do much with A and B.

But the crew weren't discouraged. If the boats couldn't be launched, they could perhaps be floated off. So they toiled on – Lightoller, Murdoch, trimmer Hemming, steward Brown, greaser Hurst, a dozen others.

On the port side Hemming struggled with the block and tackle for boat B. If he could only iron out a kink in the fall, he was sure it could still be launched. He finally got the lines working, passed the block up to Sixth Officer Moody on the roof, but Moody shouted back, 'We don't want the block; we'll leave the boat on the deck.'

Hemming saw no chance of clearing boat B this way; so he jumped and swam for it. Meanwhile the boat was pushed to the edge of the roof and slid down on some oars to the deck. It landed upside down.

On the starboard side they were having just as much trouble with boat A. Somebody propped planks against the wall of the officers' quarters, and they eased the boat down bow first. But they were still a long way from home, for the *Titanic* was now listing heavily to port, and they couldn't push the boat 'uphill' to the edge of the deck.

The men were tugging at both collapsibles when the bridge dipped under at 2.15 and the sea rolled aft along the boat deck. Colonel Gracie and Clinch Smith turned and headed for the stern. A few steps, and they were blocked by a sudden crowd of men and women pouring up from below. They all seemed to be steerage passengers.

At this moment bandmaster Hartley tapped his violin. The ragtime ended, and the strains of the Episcopal hymn

'Autumn' flowed across the deck and drifted in the still night far out over the water.

In the boats women listened with wonder. From a distance there was an agonizing stateliness about the moment. Close up, it was different. Men could hear the music, but they paid little attention. Too much was happening.

'Oh, save me! Save me!' cried a woman to Peter Daly, Lima representative of the London firm Haes and Sons, as he watched the water roll on to the deck where he stood.

'Good lady,' he answered, 'save yourself. Only God can save you now.'

But she begged him to help her make the jump, and on second thoughts he realized he couldn't shed the problem so easily. Quickly he took her by the arm and helped her overboard. As he jumped himself, a big wave came sweeping along the boat deck, washing him clear of the ship.

The sea foamed and swirled around steward Brown's feet as he sweated to get boat A to the edge of the deck. Then he realized he needn't try any longer – the boat was floating off. He jumped in . . . cut the stern lines . . . yelled for someone to free the bow . . . and in the next instant was washed out by the same wave that swept off Peter Daly.

Down, down dipped the *Titanic*'s bow, and her stern swung slowly up. She seemed to be moving forward too. It was this motion which generated the wave that hit Daly, Brown and dozens of others as it rolled aft.

Lightoller watched the wave from the roof of the officers' quarters. He saw the crowds retreating up the deck ahead of it. He saw the nimbler ones keep clear, the slower ones overtaken and engulfed. He knew that this kind of retreat just prolonged the agony. He turned and, facing the bow, dived in. As he reached the surface, he saw just ahead of him the crow's-nest, now level with the water. Blind instinct seized him, and for a moment he swam towards it as a place of safety.

Then he snapped to and tricd to swim clear of the ship. But the sea was pouring down the ventilators just in front of the forward funnel, and he was sucked back and held against the wire grating of an air shaft. He prayed it would hold. And he wondered how long he could last, pinned this way to the grating.

He never learned the answer. A blast of hot air from somewhere deep below came rushing up the ventilator and blew him to the surface. Gasping and spluttering, he finally paddled clear.

Harold Bride kept his head too. As the wave swept by, he grabbed an oarlock of collapsible B, which was still lying upside down on the boat deck near the first funnel. The boat, Bride and a dozen others were washed off together. The collapsible was still upside down, and Bride found himself struggling underneath it.

Colonel Gracie was not as sea-wise. He stayed in the crowd and jumped with the wave – it was almost like Newport. Rising on the crest, he caught the bottom rung of the iron railing on the roof of the officers' quarters. He hauled himself up and lay on his stomach right at the base of the second funnel.

Before he could rise, the roof too had dipped under.

Gracie found himself spinning round and round in a whirlpool of water. He tried to cling to the railing, then realized this was pulling him down deeper. With a mighty kick he pushed himself free and swam clear of the ship, far below the surface.

Chef John Collins couldn't do much of anything about the wave. He had a baby in his arms. For five minutes he and a deck steward had been trying to help a steerage woman with two children. First they heard there was a boat on the port side. They ran there and heard it was on the starboard side. When they got there, somebody said their best chance was to head for the stern. Bewildered, they were standing undecided – Collins holding one of the babies – when they were all swept overboard by the wave. He never saw the others again, and the child was washed out of his arms.

Jack Thayer and Milton Long saw the wave coming too. They were standing by the starboard rail opposite the second funnel, trying to keep clear of the crowds swarming towards the stern. Instead of making for a higher point, they felt the time had come to jump and swim for it. They shook hands and wished each other luck. Long put his legs over the rail, while Thayer straddled it and began unbuttoning his overcoat. Long, hanging over the side and holding the rail with his hands, looked up at Thayer and asked, 'You're coming, boy?'

'Go ahead, I'll be right with you,' Thayer reassured him.

Long slid down, facing the ship. Ten seconds later Thayer swung his other leg over the rail and sat facing out.

He was about 10 feet above the water. Then with a push he jumped out as far as he could.

Of these two techniques for abandoning ship, Thayer's was the one that worked.

The wave never reached Olaus Abelseth. Standing by the fourth funnel, he was too far back. Instead of plunging under, this part of the ship was swinging higher and higher.

As she swung up, Abelseth heard a popping and cracking . . . a series of muffled thuds . . . the crash of glassware . . . the clatter of deck chairs sliding down.

The slant of the deck grew so steep that people could no longer stand. So they fell, and Abelseth watched them slide down into the water right on the deck. Abelseth and his relatives hung on by clinging to a rope in one of the davits.

'We'd better jump or the suction will take us down,' his brother-in-law urged.

'No,' said Abelseth. 'We won't jump yet. We ain't got much show anyhow, so we might as well stay as long as we can.'

'We must jump off!' the cry came again, but Abelseth held firm: 'No, not yet.'

Minutes later, when the water was only 5 feet away, the three men finally jumped, holding each other's hands. They came spluttering to the surface, Abelseth hopelessly snarled in some rope from somewhere. He had to free his hands to untangle the line, and his cousin and brother-in-law were washed away. Somehow he got loose, but he said to himself, 'I'm a goner.'

In the maelstrom of ropes, deck chairs, planking and wildly swirling water, nobody knew what happened to most of the people. From the boats they could be seen clinging like little swarms of bees to deck houses, winches and ventilators as the stern rose higher. Close in, it was hard to see what was happening, even though – incredibly – the lights still burned, casting a sort of murky glow.

In the stories told later, Archie Butt had a dozen different endings – all gallant, none verified. According to one newspaper, Miss Marie Young, music teacher to Teddy Roosevelt's children, remembered him calling, 'Goodbye, Miss Young, remember me to the folks back home.' Yet the papers also reported Miss Young as saying she saw the iceberg an hour before the crash.

In an interview attributed to Mrs Henry B. Harris, Archie Butt was described as a pillar of strength, using his fists here – a big brother approach there – to handle the weaklings. Yet Lightoller, Gracie and the others working on the boats never saw him at all. When Mrs Walter Douglas recalled him near boat 2 around 1.45, he was standing quietly off to one side.

It was the same with John Jacob Astor. Barber August H. Weikman described his last moments with the great millionaire. It was a conversation full of the kind of small talk that normally takes place only in the barber's chair. And even more trite: 'I asked him if he minded shaking hands with me. He said, "With pleasure" . . .' Yet, barber Weikman also said he left the ship at 1.50, a good half-hour earlier.

Butt's and Astor's endings were described in a single

story attributed to Washington Dodge, the San Francisco assessor : 'They went down standing on the bridge, side by side. I could not mistake them,' the papers had him saying. Yet Dr Dodge was in boat 13, a good half-mile away.

Nor did anyone really know what happened to Captain Smith. People later said he shot himself, but there's not a shred of evidence. Just before the end steward Edward Brown saw him walk on to the bridge, still holding his megaphone. A minute later trimmer Hemmings wandered on to the bridge and found it empty. After the *Titanic* sank, fireman Harry Senior saw him in the water holding a child. Pieced together, this picture, far more than suicide, fits the kind of fighter who once said: 'In a way, a certain amount of wonder never leaves me, especially as I observe from the bridge a vessel plunging up and down in the trough of the sea, fighting her way through and over great waves. A man never outgrows that.'

Seen and unseen, the great and the unknown tumbled together in a writhing heap as the bow plunged deeper and the stern rose higher. The strains of 'Autumn' were buried in a jumble of falling musicians and instruments. The lights went out, flashed on again, went out for good. A single kerosene lantern still flickered high in the after mast.

The muffled thuds and tinkle of breaking glass grew louder. A steady roar thundered across the water as everything movable broke loose.

There has never been a mixture like it – twenty-nine boilers . . . the jewelled copy of *The Rubáiyát* . . . 800 cases of shelled walnuts . . . 15,000 bottles of ale and stout . . . huge anchor chains (each link weighed 175 pounds) . . . thirty cases of golf clubs and tennis rackets for A. G. Spalding . . . Eleanor Widener's trousseau . . . tons of coal . . . Major Peuchen's tin box . . . 30,000 fresh eggs . . . dozens of potted palms . . . five grand pianos . . . a little mantel clock in B-38 . . . the massive silver duck press.

And still it grew – tumbling trellises, ivy pots and wicker chairs in the Café Parisien . . . shuffleboard sticks . . . the fifty-phone switchboard . . . two reciprocating engines and the revolutional low-pressure turbine . . . eight dozen tennis balls for R. F. Downey & Co., a cask of china for Tiffany's, a case of gloves for Marshall Field . . . the remarkable ice-making machine on G deck . . . Billy Carter's new French Renault . . . the Ryersons' sixteen trunks, beautifully packed by Victorine.

As the tilt grew steeper, the forward funnel toppled over. It struck the water on the starboard side with a shower of sparks and a crash heard above the general uproar. Greaser Walter Hurst, struggling in the swirling sea, was half blinded by soot. He got off lucky – other swimmers were crushed under tons of steel But the falling funnel was a blessing to Lightoller, Bride and others now clinging to overturned collapsible B. It just missed the boat, washing it thirty yards clear of the plunging, twisting hull.

The *Titanic* was now absolutely perpendicular. From the third funnel after, she stuck straight up in the air, her three dripping propellers glistening even in the darkness. To Lady Duff Gordon she seemed a black finger pointing at the sky. To Harold Bride she looked like a duck that goes down for a dive.

Out in the boats, they could hardly believe their eyes. For over two hours they had watched, hoping against hope, as the *Titanic* sank lower and lower. When the water reached her red and green running lights, they knew the end was near . . . but nobody dreamed it would be like this – the unearthly din, the black hull hanging at ninety degrees, the Christmas card backdrop of brilliant stars.

Some didn't watch. In collapsible C, President Bruce Ismay bent low over his oar – he couldn't bear to see her go down. In boat 1, C. E. Henry Stengel turned his back: 'I cannot look any longer.' In No. 4, Elizabeth Eustis buried her face.

Two minutes passed, the noise finally stopped, and the *Titanic* settled back slightly at the stern. Then slowly she began sliding under, moving at a steep slant. As she glided down, she seemed to pick up speed. When the sea closed over the flagstaff on her stern, she was moving fast enough to cause a slight gulp.

'She's gone; that's the last of her,' someone sighed to lookout Lee in boat 13. 'It's gone,' Mrs Ada Clark vaguely heard somebody say in No. 4. But she was so cold she didn't pay much attention. Most of the other women were the same – they just sat dazed, dumbfounded, without showing any emotion. In No. 5, Third Officer Pitman looked at his watch and announced, 'It is 2.20.'

Ten miles away on the *Californian*, Second Officer Stone and apprentice Gibson watched the strange ship slowly disappear. She had fascinated them almost the whole watch – the way she kept firing rockets, the odd way she floated in the water. Gibson remarked that he certainly didn't think the rockets were being sent up for fun. Stone agreed: 'A ship is not going to fire rockets at sea for nothing.'

By two o'clock the stranger's light seemed very low on the horizon, and the two men felt she must be steaming away. 'Call the captain,' Stone ordered, 'and tell him that the ship is disappearing in the south-west and that she has fired altogether eight rockets.'

Gibson marched into the chart room and gave the message. Captain Lord looked up sleepily from his couch: 'Were they all white rockets?'

Gibson said yes, and Lord asked the time. Gibson replied it was 2.05 by the wheelhouse clock. Lord rolled over, and Gibson went back to the bridge.

At 2.20 Stone decided that the other ship was definitely gone, and at 2.40 he felt he ought to tell the captain himself. He called the news down the speaking tube and resumed studying the empty night.

7. 'There is Your Beautiful Nightdress Gone'

As the sea closed over the *Titanic*, Lady Cosmo Duff Gordon in boat 1 remarked to her secretary Miss Francatelli, 'There is your beautiful nightdress gone.'

A lot more than Miss Francatelli's nightgown vanished that April night. Even more than the largest liner in the world, her cargo and the lives of 1,502 people.

Never again would men fling a ship into an ice field, heedless of warnings, putting their whole trust in a few thousand tons of steel and rivets. From now on Atlantic liners took ice messages seriously, steered clear, or slowed down. Nobody believed in the 'unsinkable ship'.

Nor would icebergs any longer prowl the seas untended. After the *Titanic* sank, the American and British governments established the International Ice Patrol, and today Coast Guard cutters shepherd errant icebergs that drift towards the steamer lanes. The winter lane itself was shifted further south, as an extra precaution.

And there were no more liners with only part-time wireless. Henceforth every passenger ship had a twenty-four-hour radio watch. Never again could the world fall apart while a Cyril Evans lay sleeping off duty only ten miles away.

It was also the last time a liner put to sea without enough lifeboats. The 46,328-ton *Titanic* sailed under hopelessly outdated safety regulations. An absurd formula determined lifeboat requirements: all British vessels over 10,000 tons must carry sixteen lifeboats with a capacity of 5,500 cubic feet, plus enough rafts and floats for seventy-five per cent of the capacity of the lifeboats.

For the *Titanic* this worked out at 9,625 cubic feet. This meant she had to carry boats for only 962 people. Actually, there were boats for 1,178 – the White Star Line complained that nobody appreciated their thoughtfulness. Even so, this took care of only fifty-two per cent of the 2,207 people on board, and only thirty per cent of her total capacity. From now on the rules and formulas were simple indeed – lifeboats for everybody.

And it was the end of class distinction in filling the boats. The White Star Line always denied anything of the kind – and the investigators backed them up – yet there's overwhelming evidence that the steerage took a beating: Daniel Buckley kept from going into first class . . . Olaus Abelseth released from the poop deck as the last boat pulled away . . . steward Hart convoying two little groups of women topside, while hundreds were kept below . . . steerage passengers crawling along the crane from the well deck aft . . . others climbing vertical ladders to escape the well deck forward.

Then there were the people Colonel Gracie, Lightoller and others saw surging up from below, just before the end. Until this moment Gracie was sure the women were all off – they were so hard to find when the last boats were loading. Now, he was appalled to see dozens of them suddenly appear. The statistics suggest who they were – the

Titanic's casualty list included four of 143 first-class women (three by choice) . . . fifteen of ninety-three second-class women . . . and eighty-one of 179 third-class women.

Not to mention the children. Except for Lorraine Allison, all twenty-nine first- and second-class children were saved, but only twenty-three out of seventy-six steerage children.

Neither the chance to be chivalrous nor thc fruits of chivalry seemed to go with a third-class passage.

It was better, but not perfect, in second class. Lawrence Beesley remembered an officer stopping two ladies as they started through the gate to first class. 'May we pass to the boats?' they asked.

'No, madam; your boats are down on your own deck.'

In fairness to the White Star Line, these distinctions grew not so much from set policy as from no policy at all. At some points the crew barred the way to the boat deck; at others they opened the gates but didn't tell anyone; at a few points there were well-meaning efforts to guide the steerage up. But generally third class was left to shift for itself. A few of the more enterprising met the challenge, but most milled helplessly about their quarters – ignored, neglected, forgotten.

If the White Star Line was indifferent, so was everybody else. No one seemed to care about third class – neither the Press, the official inquiries, nor even the third-class passengers themselves.

In covering the *Titanic*, few reporters bothered to ask the third-class passengers anything. *The New York Times* was justly proud of the way it handled the disaster. Yet the famous issue covering the *Carpathia*'s arrival in New York contained only two interviews with third-class passengers. This apparently was par for the course – of forty-three survivor accounts in the *New York Herald*, two again were steerage experiences.

Certainly their experiences weren't as good copy as Lady Cosmo Duff Gordon (one New York newspaper had her saying, 'The last voice I heard was a man shouting, "My God, my God!"'). But there was indeed a story. The night was a magnificent confirmation of 'women and children first', yet somehow the loss rate was higher for third-class children than first-class men. It was a contrast which would never get by the social consciousness (or news sense) of today's Press.

Nor did Congress care what happened to third class. Senator Smith's *Titanic* investigation covered everything under the sun, including what an iceberg was made of ('Ice', explained Fifth Officer Lowe), but the steerage received little attention. Only three of the witnesses were third-class passengers. Two of these said they were kept from going to the boat deck, but the legislators didn't follow up. Again, the testimony doesn't suggest any deliberate hush-up – it was just that no one was interested.

The British Court of Inquiry was even more cavalier. Mr W. D. Harbinson, who officially represented the third-class interests, said he could find no trace of discrimination, and Lord Mersey's report gave a clean bill of health – yet not a single third-class passenger testified, and the only surviving steward stationed in steerage freely conceded that the men were kept below decks as late as 1.15 a.m.

Even the third-class passengers weren't bothered. They expected class distinction as part of the game. Olaus Abelseth, at least, regarded access to the boat deck as a privilege that went with first- and second-class passage . . . even when the ship was sinking. He was satisfied as long as they let him stay above decks.

A new age was dawning, and never since that night have third-class passengers been so philosophical.

At the opposite extreme, it was also the last time the special position of first class was accepted without question. When the White Star liner *Republic* went down in 1908, Captain Sealby told the passengers entering the lifeboats, 'Remember! Women and children go first; then the First Cabin, then the others!' There was no such rule on the *Titanic*, but the concept still existed in the public mind, and at first the Press tended to forestall any criticism over what a first-class passenger might do. When the news broke that Ismay was saved, the New York *Sun* hastened to announce, 'Ismay behaved with exceptional gallantry . . . no one knows how Mr Ismay himself got into a boat; it is assumed he wished to make a presentation of the case to his company.'

Never again would first class have it so good. In fact, almost immediately the pendulum swung the other way. Within days Ismay was pilloried; within a year a prominent survivor divorced her husband merely because, according to gossip, he happened to be saved. One of the more trying legacies left by those on the *Titanic* has been a new standard of conduct for measuring the behaviour of prominent people under stress.

It was easier in the old days . . . for the *Titanic* was also the last stand of wealth and society in the centre of public affection. In 1912 there were no movie, radio or television stars; sports figures were still beyond the pale; and café society was completely unknown. The public depended on socially prominent people for all the vicarious glamour that enriches drab lives.

This preoccupation was fully appreciated by the Press. When the *Titanic* sailed, the *New York Times* listed the prominent passengers on the front page. After she sank, the New York *American* broke the news on 16 April with a leader devoted almost entirely to John Jacob Astor; at the end it mentioned that 1,800 others were also lost.

In the same mood, the 18 April New York *Sun* covered the insurance angle of the disaster. Most of the story concerned Mrs Widener's pearls.

Never again did established wealth occupy people's minds so thoroughly. On the other hand, never again was wealth so spectacular. John Jacob Astor thought nothing of shelling out 800 dollars for a lace jacket some dealer displayed on deck when the *Titanic* stopped briefly at Queenstown. To the Ryersons there was nothing unusual about travelling with sixteen trunks. The 190 families in first class were attended by twenty-three handmaids, eight valets and assorted nurses and governesses – entirely apart from hundreds of stewards and stewardesses. These personal servants had their own lounge on C deck, so that no one need suffer the embarrassment of striking up a conversation with some handsome stranger, only to find he was Henry Sleeper Harper's dragoman.

Or take the survivors' arrival in New York. Mrs Astor was met by two automobiles, carrying two doctors, a trained nurse, a secretary and Vincent Astor. Mrs George Widener was met not by automobile but by a special train – consisting of a private Pullman, another car for ballast, and a locomotive. Mrs Charles Hays was met by a special train too, including two private cars and two coaches.

It was a reception in keeping with people who could afford as much as 4,350 dollars – and these were 1912 dollars – for a de luxe suite. A suite like this had even a private promenade deck, which figured out at something like forty dollars a front foot for six days.

This kind of life, of course, wasn't open to everybody – in fact it would take Harold Bride, who made twenty dollars a month, eighteen years to earn enough to cross in style – so those who enjoyed it gradually became part of a remarkably tightly knit little group, which also seemed to vanish with the *Titanic*.

There was a wonderful intimacy about this little world of Edwardian rich. There was no flicker of surprise when they bumped into each other, whether at the Pyramids (a great favourite), the Cowes Regatta, or the springs at Baden-Baden. They seemed to get the same ideas at the same time, and one of these ideas was to make the maiden voyage of the largest ship in the world.

So the *Titanic*'s trip was more like a reunion than an ocean passage. It fascinated Mrs Henry B. Harris, wife of the theatrical producer, who certainly wasn't part of this world. Twenty years later she still recalled with awe, 'There was a spirit of camaraderie unlike any I had experienced on previous trips. No one consulted the passenger list, to judge from the air of good fellowship that prevailed among the cabin passengers. They met on deck as one big party.'

This group knew the crew almost as well as each other. It was the custom to cross with certain captains rather than on particular ships, and Captain Smith had a personal following which made him invaluable to the White Star Line. The captain repaid the patronage with little favours and privileges which kept them coming. On the last night John Jacob Astor got the bad news direct from Captain Smith before the general alarm, and others learned too.

But the other end of the bargain was to respect the privilege. Nobody took advantage of the captain's confidence – hardly a man in the group was saved.

The stewards and waiters were on equally close terms with the group. They had often looked after the same passengers. They knew just what they wanted and how they liked things done. Every evening steward Samuel Etches would enter A-36 and lay out Thomas Andrews' dress clothes just the way Mr Andrews liked. Then at 6.45 he would return and help Andrews dress. It happened all over the ship.

And when the *Titanic* was going down, it was with genuine affection that steward Etches made Mr Guggenheim wear his sweater . . . that steward Crawford laced Mr Stewart's shoes . . . that second steward Dodd tipped off John B. Thayer that his wife was still on board, long after Thayer thought she had left. In the same spirit of devotion, dining-room steward Ray pushed Washington Dodge into boat 13 – he had persuaded the Dodges to take the *Titanic* and now felt he had to see them through.

The group repaid this loyalty with an intimacy and affection they gave few of their less-known fellow passengers. In the *Titanic*'s last hours men like Ben Guggenheim and Martin Rothschild seemed to see more of their stewards than the other passengers.

The *Titanic* somehow lowered the curtain on this way of living. It never was the same again. First the war, then income tax, made sure of that.

With this lost world went some of its prejudices – especially a firm and loudly voiced opinion of the superiority of Anglo-Saxon courage. To the survivors all stowaways in the lifeboats were 'Chinese' or 'Japanese'; all who jumped from the deck were 'Armenians', 'Frenchmen' or 'Italians'.

'There were various men passengers,' declared steward Crowe at the US inquiry, 'probably Italians, or some foreign nationality other than English or American, who attempted to rush the boats.' Steward Crowe, of course, never heard the culprits speak and had no way of knowing who they were. At the inquiry things finally grew so bad that the Italian ambassador demanded and got an apology from Fifth Officer Lowe for using 'Italian' as a sort of synonym for 'coward'.

In contrast, Anglo-Saxon blood could do no wrong. When Bride described the stoker's attack on Phillips, some newspapers made the stoker a Negro for better effect. And in a story headlined, 'Desirable Immigrants

Lost', the New York *Sun* pointed out that, along with the others, seventy-eight Finns were lost who might have done the country some good.

But along with the prejudices, some nobler instincts also were lost. Men would go on being brave, but never again would they be brave in quite the same way. These men on the *Titanic* had a touch – there was something about Ben Guggenheim changing to evening dress ... about Howard Case flicking his cigarette as he waved to Mrs Graham ... or even about Colonel Gracie panting along the decks, gallantly if ineffectually searching for Mrs Candee. Today nobody could carry off these little gestures of chivalry, but they did that night.

An air of *noblesse oblige* has vanished too. During the agonizing days of uncertainty in New York, the Astors, the Guggenheims and others like them were not content to sit by their phones or to send friends and retainers to the White Star Line offices. They went themselves. Not because it was the best way to get information, but because they felt they ought to be there in person.

Today families are as loyal as ever, but the phone would probably do. Few would insist on going themselves and braving the bedlam of the steamship office. Yet the others didn't hesitate a minute. True, Vincent Astor did get better information than the rest – and some even spoke to General Manager Franklin himself – but the point is that these people didn't merely keep in touch – they were *there*.

Overriding everything else, the *Titanic* also marked the end of a general feeling of confidence. Until then men felt they had found the answer to a steady, orderly, civilized life. For 100 years the Western world had been at peace. For 100 years technology had steadily improved. For 100 years the benefits of peace and industry seemed to be filtering satisfactorily through society. In retrospect, there may seem fewer grounds for confidence, but at the time most articulate people felt life was all right.

The *Titanic* woke them up. Never again would they be quite so sure of themselves. In technology especially, the disaster was a terrible blow. Here was the 'unsinkable ship' – perhaps man's greatest engineering achievement – going down the first time it sailed.

But it went beyond that. If this supreme achievement was so terribly fragile, what about everything else? If wealth meant so little on this cold April night, did it mean so much the rest of the year? Scores of ministers preached that the *Titanic* was a heaven-sent lesson to awaken people from their complacency, to punish them for top-heavy faith in material progress. If it was a lesson, it worked – people have never been sure of anything since.

The unending sequence of disillusionment that has followed can't be blamed on the *Titanic*, but she was the first jar. Before the *Titanic*, all was quiet. Afterwards, all was tumult. That is why, to anybody who lived at the time, the *Titanic* more than any other single event marks the end of the old days, and the beginning of a new, uneasy era.

There was no time for such thoughts at 2.20 a.m., Monday 15 April 1912. Over the *Titanic*'s grave hung a thin, smoky vapour, soiling the clear night. The glassy sea was littered with crates, deck chairs, planking, pilasters and cork-like rubbish that kept bobbing to the surface from somewhere now far below.

Hundreds of swimmers thrashed the water, clinging to the wreckage and each other. Steward Edward Brown, gasping for breath, dimly noticed a man tearing at his clothes. Third-class passenger Olaus Abelseth felt a man's arm clamp around his neck. Somehow he wriggled loose, spluttering, 'Let go!' But the man grabbed him again, and it took a vigorous kick to free himself for good.

If it wasn't the people, it was the sea itself that broke a man's resistance. The temperature of the water was twenty-eight degrees – well below freezing. To Second Officer Lightoller it felt like 'a thousand knives' driven into his body. In water like this, lifebelts did no good.

Yet a few dozen managed to keep both their wits and their stamina. For these, two hopes of safety loomed in the littered water – collapsibles A and B. Both had floated off the sinking boat deck, A swamped and B upside down. Then the falling funnel washed both boats further clear of the crowd. Now the strongest and luckiest swimmers converged upon them.

After about twenty minutes Olaus Abelseth splashed alongside A. Perhaps a dozen others already lay half-dead in the wallowing boat. They neither helped nor hindered him as he scrambled over the gunwale. They just mumbled, 'Don't capsize the boat.'

One by one others arrived, until some two dozen people slumped in the hulk. They were a weird assortment – tennis star R. Norris Williams II, lying beside his waterlogged fur coat . . . a couple of Swedes . . . fireman

John Thompson with badly burned hands . . . a first-class passenger in underpants . . . steward Edward Brown . . . third-class passenger Mrs Rosa Abbott.

Gradually boat A drifted further away; the swimmers arrived at less frequent intervals. Finally they stopped coming altogether, and the half-swamped boat drifted silent and alone in the empty night.

Meanwhile other swimmers were making for overturned collapsible B. This boat was much closer to the scene. Many more people swarmed around its curved white keel, and they were much louder, much more active.

'Save one life! Save one life!' Walter Hurst heard the cry again and again as he joined the men trying to board the collapsible.

Wireless operator Harold Bride was of course there from the start, but under the boat. Lightoller also arrived before the *Titanic* sank. He was treading water alongside when the forward funnel fell. The wave almost washed him away and at the same time pushed young Jack Thayer right up against the boat. By now Hurst and three or four others were crouching on the keel. Lightoller and Thayer scrambled aboard too. Bride was still under the boat, lying on his back, bumping his head against the seats, gulping for air in the stuffy darkness.

Then came A. H. Barkworth, a Yorkshire Justice of the Peace. He wore a great fur coat over his lifebelt, and this daring arrangement surprisingly helped buoy him up. Fur coat and all, he too clambered on to the upturned collapsible, like some bedraggled, shaggy animal.

Colonel Gracie arrived later. Dragged down with the

Titanic, he tried first a plank, then a large wooden crate, before he spied the overturned collapsible. When he finally drew alongside, more than a dozen men were lying and kneeling on the bottom.

No one offered a helping hand. With each new man, the collapsible sagged lower into the sea; already the water slopped over the keel from time to time. But Gracie hadn't come this far for nothing. He grabbed the arm of a man already lying on the boat, and hauled himself on to the keel. Next, assistant cook John Collins swam up and managed to get on too. Then Bride dived out from underneath and scrambled on to the stern.

By the time steward Thomas Whiteley arrived, collapsible B wallowed under the weight of thirty men. As he tried to climb aboard, someone swatted him with an oar, but he made it anyhow. Fireman Harry Senior was beaten off by an oar, but he swam around to the other side and finally persuaded them to let him on too.

All the time men straddling the stern and the bow flayed the water with loose boards, paddling to get away from the scene and steer clear of the swimmers.

'Hold on to what you have, old boy. One more would sink us all,' the men in the boat shouted to those in the water.

'That's all right, boys; keep cool,' one of the swimmers replied when they asked him to stay clear. Then he swam off, calling back, 'Good luck, God bless you.'

Another swimmer kept cheering them on: 'Good boy; good lads!' He had the voice of authority and never asked to climb aboard. Even though they were dangerously overcrowded, Walter Hurst couldn't resist holding out an oar. But the man was too far gone. As the oar touched him, he spun about like a cork and was silent. To this day Hurst thinks it was Captain Smith.

As they moved off into the lonely night, away from the wreckage and the swimmers, one of the seamen lying on the keel hesitantly asked, 'Don't the rest of you think we ought to pray?'

Everybody agreed. A quick poll showed Catholics, Presbyterians, Episcopalians, Methodists all jumbled together; so they compromised on the Lord's Prayer, calling it out in chorus with the man who suggested it as their leader.

It was not the only sound that drifted over the water. All the time while collapsibles A and B were filling up and painfully struggling away from the scene, hundreds of swimmers were crying for help. Individual voices were lost in a steady, overwhelming clamour. To fireman George Kemish, tugging at his oar in boat 9, it sounded like a hundred thousand fans at a British football cup final. To Jack Thayer, lying on the keel of boat B, it seemed like the high-pitched hum of locusts on a midsummer night in the woods back home in Pennsylvania.

8. 'It Reminds Me of a Bloomin' Picnic'

The cries in the night meant one thing to lively, impulsive Fifth Officer Lowe – row back and help.

He was in a good position to do something. After leaving the *Titanic* in No. 14, he had rounded up boats 10, 12, 4 and D, and all five were tied together like a string of beads 150 yards away.

'Consider yourselves under my command,' he ordered, and now he organized his flotilla for rescue work. It was suicide for all the boats to go – they were too undermanned to face the bedlam – but one boat with a hand-picked crew might do some good. So Lowe divided his fifty-five passengers among the other four boats and picked volunteers from each to give No. 14 some expert oarsmen.

It was nerve-racking work, playing musical chairs with rowboats at 2.30 a.m. in the middle of the Atlantic – almost more than Lowe could stand. 'Jump, God damn you, jump!' he shouted impatiently at Miss Daisy Minahan. On the other hand, an old lady in a shawl seemed much too agile; Lowe ripped it off and looked into the frightened face of a young man – eyes white with terror. This time nothing was said, but he pitched the man into boat No. 10 as hard as he could.

It took time to make the transfer. Then more time

while Lowe waited for the swimmers to thin out enough to make the expedition safe. Then still more time to get there. It was after three o'clock – nearly an hour since the sinking – when boat 14 edged into the wreckage and the people.

There was little left – steward John Stewart . . . first-class passenger W. F. Hoyt . . . a Japanese steerage passenger, who had lashed himself to a door. For nearly an hour boat 14 played a hopeless blind-man's-buff, chasing after shouts and calls in the darkness, never quite able to reach whoever was shouting.

They got only four, and Mr Hoyt died in an hour. Lowe had miscalculated how long it took to row to the scene . . . how long to locate a voice in the dark . . . most of all, how long a man could live in water at twenty-eight degrees. There was, he learned, no need to wait until the crowd 'thinned out'. But at least Lowe went back.

Third Officer Pitman in No. 5 also heard the cries. He turned the boat around and shouted, 'Now, men, we will pull towards the wreck!'

'Appeal to the officer not to go back,' a lady begged steward Etches as he tugged at his oar. 'Why should we lose all our lives in a useless attempt to save others from the ship?'

Other women protested too. Pitman was torn by the dilemma. Finally he reversed his orders and told his men to lay on their oars. For the next hour No. 5 – forty people in a boat that held sixty-five – heaved gently in the calm Atlantic swell, while its passengers listened to the swimmers three hundred yards away.

In No. 2, Steward Johnson recalled Fourth Officer Boxhall asking the ladies, 'Shall we go back?' They said no; so boat 2 – about sixty per cent full – also drifted while her people listened.

The ladies in boat 6 were different. Mrs Lucien Smith, stung that she had fallen for the white lie her husband used to get her in the boat; Mrs Churchill Candee, moved by the gallantry of her self-appointed protectors; Mrs J. J. Brown, naturally brave and lusty for adventure – all begged Quartermaster Hitchens to return to the scene. Hitchens refused. He painted a vivid picture of swimmers grappling at the boat, of No. 6 swamping and capsizing. The women still pleaded, while the cries grew fainter. Boat No. 6 – capacity sixty-five; occupants twenty-eight – went no closer to the scene.

In No. 1, fireman Charles Hendrickson sang out, 'It's up to us to go back and pick up anyone in the water.'

Nobody answered. Lookout George Symons, in charge of the boat, made no move. Then, when the suggestion came again, Sir Cosmo Duff Gordon announced he didn't think they should try; it would be dangerous; the boat would be swamped. With that, the subject was dropped. No. 1 – twelve people in a boat made for forty – rowed on aimlessly in the dark.

In boat after boat the story was the same: a timid suggestion, a stronger refusal, nothing done. Of 1,600 people who went down on the *Titanic*, only thirteen were picked up by the eighteen boats that hovered nearby. Boat D hauled in Mr Frederick Hoyt because he planned it that way. Boat 4 rescued eight – not because it rowed back but because it was within reach. Only No. 14 returned to the scene. Why the others didn't is part of the mystery of why trained men in identical situations should react so differently.

As the cries died away, the night became strangely peaceful. The *Titanic*, the agonizing suspense, was gone. The shock of what had happened, the confusion and excitement ahead, the realization that close friends were lost for ever had not yet sunk in. A curiously tranquil feeling came over many of those in the boats.

With the feeling of calm came loneliness. Lawrence Beesley wondered why the *Titanic*, even when mortally wounded, gave everyone a feeling of companionship and security that no lifeboat could replace. In No. 3, Elizabeth Shutes watched the shooting stars and thought to herself how insignificant the *Titanic*'s rockets must have looked, competing against nature. She tried to bury her loneliness by pretending she was back in Japan. Twice she had left there at night too, lonely and afraid, but everything came out all right in the end.

In boat 4, Miss Jean Gertrude Hippach also watched the shooting stars – she had never seen so many. She recalled a legend that every time there's a shooting star, somebody dies.

Slowly – very slowly – life in the boats picked up again. Fourth Officer Boxhall started firing off green flares from boat 2. Somehow this brought people out of their trance, cheered them up too. It was hard to judge distance, and some thought the green flares were being fired by rescue ships on the horizon.

14. The iceberg that sank the *Titanic*? It was photographed near the scene on 15 April by the Chief Steward of the German ship *Prinz Adalbert*. He took it not because of the *Titanic* – the news had not yet reached him – but because a great red scar of red paint ran along the iceberg's base, suggesting a recent collision with some ship (author)

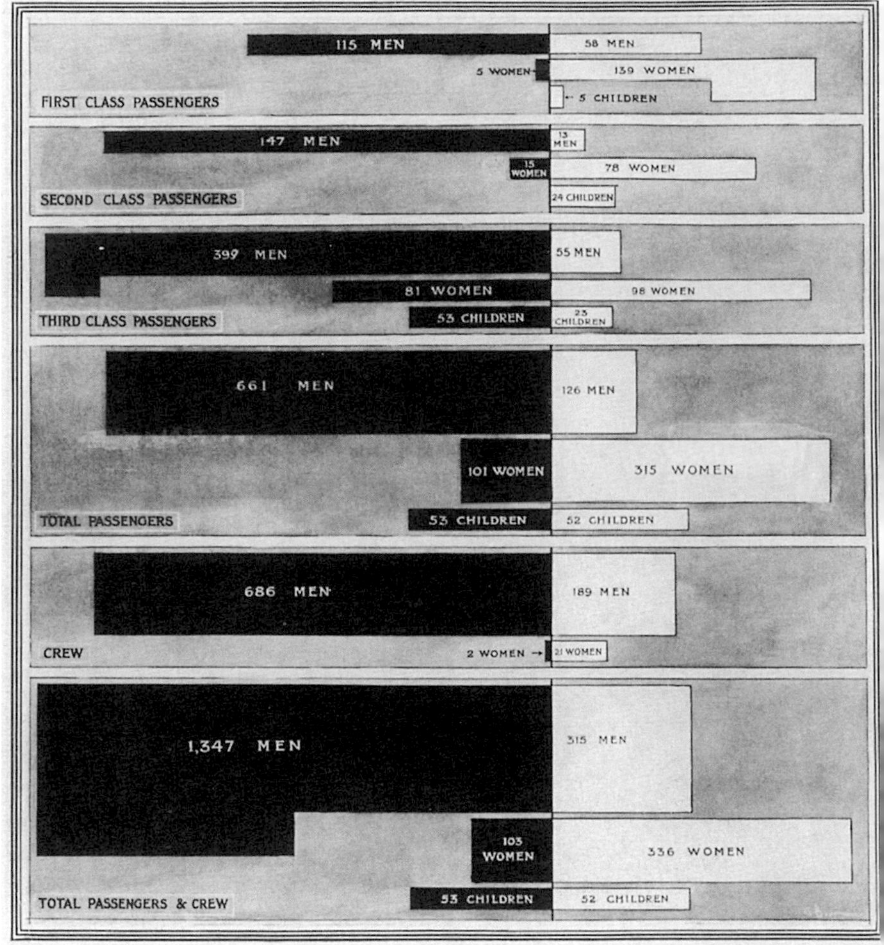

15. Statistics told only one part of the tragedy. Third-class casualties included the entire Sage family of eleven and all five Rice children (*Illustrated London News*)

16. No matter how valid the circumstances, any male survivor from first class came under severe scrutiny – note the quotes around 'men' in the caption of this cartoon from the London *Daily Herald* (photo John Webb)

17. *Left*, controversy swirled around Sir Cosmo Duff Gordon, who left the *Titanic* in boat 1, carrying only twelve people. *Right*, Lady Duff Gordon. She and her secretary were the only women in boat 1 (both *Illustrated London News*)

Fifth Officer Lowe rescued swimmers from the water

The Countess of Rothes handled the tiller of boat 8

Captain Edward Smith, last seen swimming in the debris, holding a child

Thomas Andrews helped many women escape from the ship he built

Baker Charles Joughin was probably the last man off

18. *Titanic* heroes (*Illustrated London News*)

19. When the crisis came, the low-paid and hard-worked crew set a matchless example of devotion to duty. The ship's band, pictured above, played on with ragtime until the water was over their feet; all were lost (*Illustrated London News*)

20. All day on 15 April, anxious crowds besieged the White Star offices in New York. They were assured that the *Titanic* was practically unsinkable (Brown Brothers)

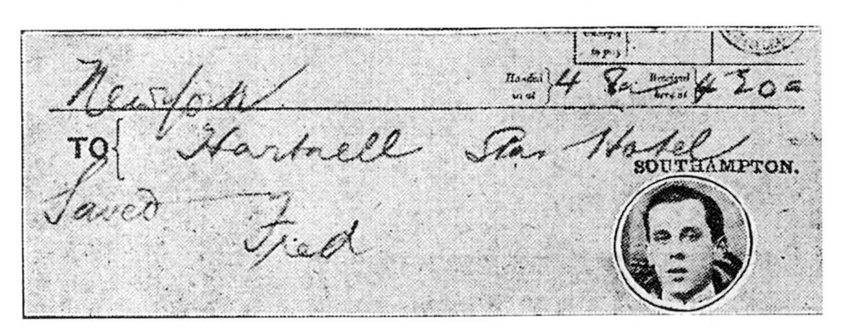

New York

TO Hartnell Star Hotel SOUTHAMPTON.

Saved

Fred

21. A one-word telegram told steward Fred Hartnell's family all they wanted to know (photo John Webb)

22. Southampton crowds scan the lists of lost and saved posted outside the White Star offices. Many of the crew lived here; in one street twenty families were bereaved (*Illustrated London News*)

THE WRECK of the TITANIC

Descriptive Musical Sketch for the PIANO

by HAYDON AUGARDE.

45*
THE LAWRENCE WRIGHT MUSIC Co
Copyright.
8, DENMARK STREET, (CHARING CROSS ROAD.)
PRINTED IN ENGLAND.
LONDON, W.C.
ALSO AT LEICESTER.

Agents for Australia and New Zealand.
ALLAN & Co. Proprietory Ltd.
276-278, Collins St.
MELBOURNE.

23. Memorial pictures, heavily bordered in black, were eagerly bought from street vendors, as were souvenir napkins, postcards, buttons and bits of pottery. Columns of incredibly poor poetry emerged, and at last eight different pieces of commemorative sheet music appeared on the stands (author)

Oars squealed and splashed in the water, voices sang out as the boats hailed one another in the dark. Nos. 5 and 7 tied up together; so did Nos. 6 and 16. Boat 6 borrowed a stoker for extra rowing power. Other boats were drifting apart. Over a radius of four or five miles, eighteen little boats wandered about through the night or drifted together on a sea flat as a reservoir. A stoker in No. 13 thought of times he had spent on the Regent's Park lake and blurted out, 'It reminds me of a bloomin' picnic!'

At times it did seem like a picnic – the small talk, the children under foot. Lawrence Beesley tried to tuck a blanket under the toes of a crying baby, and discovered that he and the lady holding the baby had close mutual friends in Clonmel, Ireland. Edith Russell amused another baby with her toy pig that played the 'Maxixe' whenever its tail was twisted. Hugh Woolner found himself feeding cookies to four-year-old Louis Navatril. Mrs John Jacob Astor lent a steerage woman a shawl to comfort her little daughter whimpering from the cold. The woman thanked Mrs Astor in Swedish, and wrapped the shawl around her little girl.

About this time Marguerite Frolicher was introduced to an important piece of picnic equipment. Still deathly seasick, she was noticed by a kindly gentleman sitting nearby. He pulled out a silver flask with a cup top and suggested a drink of brandy might help. She took the suggestion and was instantly cured. Perhaps it was the brandy, perhaps the novel experience – in all her twenty-two years, she had never seen a flask before, and she was fascinated.

But no picnic was ever so cold. Mrs Crosby shivered so hard in No. 5 that Third Officer Pitman wrapped a sail around her. A stoker in No. 6 sat beside Mrs Brown, his teeth chattering with the cold. Finally she wrapped her sable stole about his legs, tying the tails around his ankles. In No. 16 a man in white pyjamas looked so cold he reminded the other passengers of a snowman. Mrs Charlotte Collyer was so numb she toppled over in No. 14; her hair caught in an oarlock and a big tuft came out by the roots.

The crew did their best to make the women more comfortable. In No. 5 a sailor took off his stockings and gave them to Mrs Washington Dodge. When she looked up in startled gratitude, he explained, 'I assure you, ma'am, they are perfectly clean, I just put them on this morning.'

In No. 13, fireman Beauchamp shivered in his thin jumpers, but he refused to take an extra coat offered him by an elderly lady, insisting it go to a young Irish girl instead. For the people in this boat there was additional relief from an unexpected quarter. When steward Ray left his cabin for the last time, he picked up six handkerchiefs lying in his trunk. Now he gave them out, telling people to tie a knot in each corner and turn them into caps. As a result, he recalls with pride, 'six heads were crowned'.

Besides the cold, the number of lady oarsmen dispelled the picnic illusion. In No. 4, Mrs John B. Thayer rowed for five hours in water up to her shins. In No. 6 the irrepressible Mrs Brown organized the women, two to an oar. One held the oar in place, while the second did the pulling. In this way Mrs Brown, Mrs Meyer, Mrs Candee and others propelled the boat some three or four miles, in a hopeless effort to overtake the light that twinkled on the horizon most of the night.

Mrs Walter Douglas handled the tiller of boat 2. Boxhall, who was in charge, pulled an oar and helped to fire the green flares. Mrs J. Stuart White didn't help to row No. 8, but she appointed herself a sort of signalman. She had a cane with a built-in electric light, and during most of the night she waved it fiercely about, alternately helping and confusing everyone.

In No. 8, Marie Young, Gladys Cherry, Mrs F. Joel Swift and others pulled at the oars. Mrs William R. Bucknell noted with pride that as she rowed next to the Countess of Rothes, further down the boat her maid was rowing next to the countess's maid. Most of the night the countess handled the tiller. Seaman Jones, in charge, later explained to the *Sphere* why he put her there: 'There was a woman in my boat as was a woman . . . When I saw the way she was carrying herself and heard the quiet determined way she spoke to the others, I knew she was more of a man than any we had on board.' At the American investigation, perhaps lacking the guidance of the Press, Jones phrased it a little less elegantly: 'She had a lot to say, so I put her to steering the boat.'

But there was no doubt how he felt. After the rescue Jones removed the numeral '8' from the lifeboat, had it framed, and sent it to the countess to show his admiration.

As the night wore on, the early composure began to give way. In No. 3, Mrs Charles Hays hailed the boats that came near, searching for her husband. 'Charles Hays, are you there?' she would call over and over again. In No. 8 Señora de Satode Penasco screamed for her husband Victor, until the Countess of Rothes couldn't stand it any longer. Turning the tiller over to her cousin Gladys Cherry, she slipped down beside the señora and spent the rest of the night rowing beside her, trying to cheer her up. In No. 6, Madamc dc Villiers constantly called for her son, who wasn't even on the *Titanic.*

Gradually a good deal of squabbling broke out. The women in No. 3 bickered about trifles, while their husbands sat in embarrassed silence. Mrs Washington Dodge – who wanted to row back against the wishes of nearly everyone else in No. 5 – grew so bitter that when No. 7 came by, she switched boats in mid-ocean. Maud Slocombe, the *Titanic*'s irrepressible masseuse, helped to bawl out a woman who kept setting off, of all things, an alarm clock in No. 11. Seaman Diamond, a tough ex-boxer in charge of No. 15, swore oaths that turned the night air even bluer.

Many of the arguments revolved around smoking. In 1912 tobacco was not yet the American cure-all for easing strain and tension, and the women in the boats were shocked. Miss Elizabeth Shutes begged two men sitting near her in No. 3 to stop smoking, but they kept on. To Mrs J. Stuart White it was a matter that still rankled at the investigation. When Senator Smith asked if she wished to mention anything bearing on the crew's discipline, she exploded, 'As we cut loose from the ship these stewards took out cigarettes and lighted them. On an occasion like that!'

In the more casual intimacy of boat 1, smoking was no problem. When Sir Cosmo Duff Gordon gave fireman Hendrickson a good cigar, neither of the women in the boat could very well object. Miss Francatelli was employed by Sir Cosmo's wife, and Lady Duff Gordon was too sick to care. With her head down upon the oars and tackle, she vomited away the night.

But No. 1 had its squabbles too. Sir Cosmo and Mr C. E. Henry Stengel of Newark, New Jersey, didn't get along very well. This might not have mattered in a crowded boat, but with only twelve people it was rather grating. According to Sir Cosmo, Mr Stengel kept shouting, 'Boat ahoy!' He also gave lookout Symons conflicting advice on where to steer. Nobody paid any attention, but he irritated Sir Cosmo so much that he finally asked Mr Stengel to keep quiet. Sir Cosmo was doubly annoyed when Mr Stengel later testified that 'between Sir Cosmo and myself we decided which way to go'.

Meanwhile fireman Pusey was smouldering over Lady Duff Gordon's remark to Miss Francatelli on the loss of her nightgown. At the time he told her, 'Never mind, you have saved your lives; but we have lost our kit.'

Half an hour later, still smouldering, Pusey turned to Sir Cosmo, 'I suppose you have lost everything?'

'Of course.'

'But you can get more?'

'Yes.'

'Well, we have lost our kit, and the company won't give us any more. And what's more, our pay stops from tonight.'

Sir Cosmo had had enough: 'Very well, I will give you a fiver each to start a new kit.'

He did, too, but lived to regret it. The Duff Gordons' near monopoly of boat 1, its failure to row back, gave the gift the look of a payoff that Sir Cosmo had a hard time living down.

Nor did subsequent events help his case. When Lady Duff Gordon reassembled the men in life jackets for a group picture after the rescue, they looked more and more like the Duff Gordons' personal crew. Still later, when it came out that lookout Symons, nominally in charge of the boat, spent the day with Sir Cosmo's lawyer just before he testified at the British inquiry, it looked as though Sir Cosmo even had his personal coxswain.

There's no evidence that Sir Cosmo was guilty of more than extreme bad taste.

Drinking caused some trouble too. When No. 4 plucked a member of the crew out of the water, he had a bottle of brandy in his pocket. It was then thrown away because, as he explained it a few weeks later to the Press, 'it was feared that if any hysterical person in the boat touched it, the result might be bad'. Miss Eustis has a somewhat different version: 'One man was drunk and had a bottle of brandy in his pocket, which the quartermaster promptly threw overboard, and the drunken man was thrown into the bottom of the boat . . .'

There was a different kind of trouble in boat 6. Friction erupted from the moment Major Peuchen slid down the line to fill out the crew. Peuchen, used to giving orders, couldn't resist trying to take command. Quartermaster

Hitchens had other ideas. As they rowed away from the *Titanic*, Peuchen was pulling an oar and Hitchens was at the tiller, but within ten minutes Peuchen asked Hitchens to let a lady steer and join him in rowing. The Quartermaster answered that he was in charge and Peuchen's job was to row and keep quiet.

Painfully the boat struggled away, with just Peuchen and lookout Fleet rowing. Under Mrs Brown's leadership, most of the women gradually joined in, but Hitchens remained glued to the tiller, shouting at them to row harder or all would be sucked under when the *Titanic* sank.

Then women began shouting back, and as the boat splashed along in the dark, the night echoed with bitter repartee. Most of the time No. 6 was heading towards the elusive light on the horizon, and when it became clear they could never reach it, Hitchens announced that all was lost; they had neither water nor food nor compass nor charts; they were hundreds of miles from land and didn't even know which way they were heading.

Major Peuchen had given up by now, but the women tore into him. Mrs Candee grimly showed him the North Star. Mrs Brown told him to shut up and row. Mrs Meyer jeered at his courage.

Then they tied up with boat 16, and Hitchens ordered them to drift. But the women couldn't stand the cold and insisted on rowing to keep warm. Mrs Brown gave an oar to a grimy stoker transferred from No. 16, and told everyone to row. Hitchens moved to stop her, and Mrs Brown told him if he came any closer, she would throw him overboard.

He sank back under a blanket and began shouting insults. Mrs Meyer answered back – accused him of taking all the blankets and drinking all the whisky. Hitchens hotly denied it. The transferred stoker, wondering what on earth he had run into, called out, 'I say, don't you know you're talking to a lady.'

Hitchens yelled back, 'I know whom I'm speaking to, and I'm commanding this boat!'

But the stoker's rebuke worked. The quartermaster lapsed into silence. Boat 6 rowed on through the night with Hitchens subdued, Peuchen out of the picture, and Mrs Brown virtually in charge.

Even among the men clinging desperately to overturned boat B, there was time for petty bickering. Colonel Gracie – his teeth chattering, his matted hair now frozen stiff – noticed the man beside him wore a dry outing cap. The colonel asked to borrow the cap to warm his head for a minute. 'And what would I do?' the man shot back.

Nerves on boat B were understandably frayed. The air was leaking out from the hull, and every minute it sank a little lower in the water. The sea occasionally washed over the keel, and one impulsive move might pitch everybody into the sea. They needed cool leadership badly.

At this point Gracie was relieved to hear the deep, rich voice of Second Officer Lightoller, and even more relieved when a somewhat tipsy crewman on the boat called out, 'We will all obey what the officer orders.'

Lightoller quickly responded. Feeling that only concerted, organized action would keep the boat balanced, he had all thirty men stand up. He arranged them in a double column, facing the bow. Then as the boat lurched with the sea, he shouted, 'Lean to the right' . . . 'Stand upright' . . . 'Lean to the left' – whatever was necessary to counteract the swell.

As they threw their weight this way and that, for a while they yelled, 'Boat ahoy! Boat ahoy!' Lightoller finally stopped them, urging them to save their strength.

It grew still colder, and the colonel complained again about his head, this time to Lightoller. Another man offered them both a pull from his flask. They turned him down but pointed out Walter Hurst, shivering nearby. Hurst thought it was brandy and took a big swig. He almost choked – it was essence of peppermint.

They talked a surprising amount. Assistant cook John Maynard told how Captain Smith swam alongside the boat just before the *Titanic* took her last plunge. They pulled him on, but he slipped off again. Later fireman Harry Senior claimed the captain let go on purpose, saying 'I will follow the ship!' It may have been true, but Hurst is sure the captain never reached the boat. Besides, Senior was one of the last to arrive – probably too late to have seen the captain himself.

Most of all they all talked of getting rescued. Lightoller soon discovered Harold Bride, the junior wireless operator, at the stern of the boat, and from his position in the bow he asked what ships were on the way. Bride shouted back: the *Baltic*, the *Olympic*, the *Carpathia*. Lightoller figured the *Carpathia* should arrive at daybreak . . . passed the word around, to buck up sagging spirits.

From then on they scanned the horizon searching for

any sign. From time to time they were cheered by the green flares lit by Boxhall in boat 2. Even Lightoller thought they must come from another ship.

Slowly the night passed. Towards dawn a slight breeze sprang up. The air seemed even more frigid. The sea grew choppy. Bitter-cold waves splashed over the feet, the shins, the knees of the men on boat B. The spray stabbed their bodies and blinded their eyes. One man, then another, then another rolled off the stern and disappeared from sight. The rest fell silent, completely absorbed in the battle to stay alive.

The sea was silent too. No one saw a trace of life in the waves that rippled the smooth Atlantic as the first light of dawn streaked the sky.

But one man still lived – thanks to a remarkable combination of initiative, luck and alcohol. Four hours earlier, chief baker Charles Joughin was awakened, like so many on the *Titanic*, by that strange, grinding jar. And like many others he heard the call to general quarters a little after midnight.

But Joughin didn't merely report to the boat deck. He reasoned that if boats were needed, provisions were needed too; so on his own initiative he mustered his staff of thirteen bakers and ransacked the *Titanic*'s larder of all spare bread. The bakers then trooped topside carrying four loaves apiece.

This done, Joughin retired to his cabin on E deck, port side, for a nip of whisky.

About 12.30 he felt sufficiently fortified to reascend the stairs to his boat, No. 10. At this stage it was still difficult to persuade the women to go; so Joughin resorted to stronger methods. He went down to the promenade deck and hauled some up by force. Then, to use his own word, he 'threw' them into the boat. Rough but effective.

Joughin was assigned to No. 10 as skipper, but he thought there were enough men to handle the boat; so he jumped out and helped launch it instead. To go with it, he explained, 'would have set a bad example'.

It was now 1.20. He scampered down the slanting stairs again to his cabin on E deck and poured himself another drink. He sat down on his bunk and nursed it along – aware but not particularly caring that the water now rippled through the cabin doorway, swilled across the chequered linoleum, and rose to the top of his shoes.

About 1.45 he saw, of all people, gentle old Dr O'Loughlin poking around. It never occurred to Joughin to wonder what the old gentleman was doing way down here, but the proximity of the pantry suggests that Joughin and the doctor were thinking along similar lines.

In any case, Joughin greeted him briefly, then went back up to the boat deck. None too soon, for the *Titanic* was listing heavily now, and the slant was much steeper. Any longer, and the stairs might have been impossible.

Though all the boats were gone, Joughin was anything but discouraged. He went down to B deck and began throwing deck chairs through the windows of the enclosed promenade. Others watched him, but they didn't help. Altogether he pitched about fifty chairs overboard.

It was tiring work; so after he lugged the last chair to the edge and squeezed it through the window (it was a little like threading a needle), Joughin retired to the pantry on the starboard side of A deck. It was 2.10.

As he quenched his thirst – this time it was water – he heard a kind of crash, as though something had buckled. The pantry cups and saucers flew about him, the lights glowed red, and overhead he heard the pounding of feet running aft.

He bolted out of the pantry towards the stern end of A deck, just behind a swarm of people, running the same way and clambering down from the boat deck above. He kept out of the crush as much as possible and ran along in the rear of the crowd. He vaulted down the steps to B deck, then to the well deck. Just as he got there, the *Titanic* gave a sickening twist to port, throwing most of the people into a huge heap along the port rail.

Only Joughin kept his balance. Alert but relaxed, his equilibrium was marvellous, as the stern rose higher and corkscrewed to port. The deck was now listing too steeply to stand on, and Joughin slipped over the starboard rail and stood on the actual side of the ship. He worked his way up the side, still holding on to the rail – but from the outside – until he reached the white-painted steel plates of the poop deck. He now stood on the rounded stern end of the ship, which had swung high in the air some 150 feet above the water.

Joughin casually tightened his lifebelt. Then he glanced at his watch – it said 2.15. As an afterthought, he took it off and stuck it into his hip pocket. He was beginning to puzzle over his position when he felt the stern beginning to drop under his feet – it was like taking an elevator. As the sea closed over the stern, Joughin stepped off into the water. He didn't even get his head wet.

He paddled off into the night, little bothered by the freezing water. For over an hour he bobbed about, moving his arms and legs just enough to keep upright. 'No trick at all,' he explains cheerfully today.

It was four o'clock when he saw what he thought was wreckage in the first grey light of day. He swam over and discovered it was the upturned collapsible B.

The keel was crowded and he couldn't climb on, so he hung around for a while until he spied an old friend from the kitchen – entrée chef John Maynard. Blood proved thicker than water; Maynard held out his hand and Joughin hung on, treading water, still thoroughly insulated.

The others didn't notice him . . . partly because they were too numb to care, partly because all eyes now scanned the south-east horizon. It was just after 3.30 when they first saw it – a distant flash followed by a far-off boom. In boat 6, Miss Norton cried, 'There's a flash of lightning!' while Hitchens growled, 'It's a falling star!' In No. 13 a stoker lying in the bottom, almost unconscious from the cold, bolted up, shouting, 'That was cannon!'

In No. 8, seaman Jones hardly dared believe his eyes. Turning to the Countess of Rothes, rowing next to him, he whispered, 'Can you see any lights? Look on the next wave we top, but don't say anything in case I am wrong.'

As the boat heaved up on the next swell, the countess scanned the horizon. Far off, she saw a dim light. A few moments later there was no doubt about it, and they told the others.

The single light grew brighter; then another appeared; then row after row. A big steamer was pounding up, firing rockets to reassure the *Titanic*'s people that help was on the way. In No. 9, deck hand Paddy McGough suddenly thundered, 'Let us all pray to God, for there is a ship on the horizon and it's making for us!'

The men in boat B let out a yelp of joy and started babbling again. Someone lit a newspaper in No. 3 and waved it wildly, then Mrs Davidson's straw hat – it would burn longer. In Mrs A. S. Jerwan's boat they dipped handkerchiefs in kerosene and lit them as signals. In No. 13 they twisted a paper torch out of letters. Boxhall burned a last green flare in boat 2. In No. 8, Mrs White swung her electric cane as never before.

Over the water floated cheers and yells of relief. Even nature seemed pleased, as the dreary night gave way to the mauve and coral of a beautiful dawn.

Not everyone saw it. In half-swamped boat A, Olaus Abelseth tried to kindle the will to live in a half-frozen man lying beside him. As day broke, he took the man's shoulder and raised him up, so that he was sitting on the floorboards. 'Look!' pleaded Abelseth, 'we can see a ship now; brace up!'

He took one of the man's hands and raised it. Then he shook the man's shoulder. But the man only said, 'Who are you?' And a minute later, 'Let me be . . . who are you?'

Abelseth held him up for a while; but it was such a strain, he finally had to use a board as a prop. Half an hour later the sky blazed with thrilling, warm shades of pink and gold, but now it was too late for the man to know.

9. 'We're Going North Like Hell'

Mrs Anne Crain puzzled over the cheerful smell of coffee brewing as she lay in her cabin on the Cunarder *Carpathia*, bound from New York to the Mediterranean. It was nearly 1.00 a.m. on the fourth night out, and by now Mrs Crain knew the quiet little liner well enough to feel that any sign of activity after midnight was unusual, let alone coffee brewing.

Down the corridor Miss Ann Peterson lay awake in her bunk too. She wondered why the lights were turned on all over the ship – normally the poky *Carpathia* was shut down by now.

Mr Howard M. Chapin was more worried than puzzled. He lay in the upper berth of his cabin on A deck – his face just a few inches below the boat deck above. Some time after midnight a strange sound suddenly woke him up. It was a man kneeling down on the deck, directly over his head. The day before, he had noticed a lifeboat fall tied to a cleat just about there; now he felt sure the man was unfastening the boat and something was wrong.

Nearby, Mrs Louis M. Ogden awoke to a cold cabin and a speeding ship. Hearing loud noises overhead, she too decided something must be wrong. She shook her sleeping husband. His diagnosis didn't reassure her – the noise was the crew breaking out the chocks from the lifeboats overhead. He opened the stateroom door and saw a line of stewards carrying blankets and mattresses. Not very reassuring either.

Here and there, all over the ship, the light sleepers listened restlessly to muffled commands, tramping feet, creaking davits. Some wondered about the engines – they were pounding so much harder, so much faster than usual. The mattress jiggled wildly ... the washstand tumblers rattled loudly in the brackets ... the woodwork groaned with the strain. A turn of the tap produced only cold water – a twist on the heater knob brought no results – the engines seemed to be feeding on every ounce of steam.

Strangest of all was the bitter cold. The *Carpathia* had left New York on 11 April, bound for Gibraltar, Genoa, Naples, Trieste and Fiume. Her 150 first-class passengers were mostly elderly Americans following the sun in this pre-Florida era; her 575 steerage passengers were mostly Italians and Slavs returning to their sunny Mediterranean. All of them welcomed the balmy breeze of the Gulf Stream that Sunday afternoon. Towards five o'clock it grew so warm that Mr Chapin shifted his deck chair to the shade. Now there was an amazing change – the frigid blast that swept through every crack and seam felt like the Arctic.

On the bridge, Captain Arthur H. Rostron wondered whether he had overlooked anything. He had been at sea for twenty-seven years – with Cunard for seventeen – but this was only his second year as a Cunard skipper and only his third month on the *Carpathia*. The *Titanic*'s call for help was his first real test.

When the CQD arrived, Rostron had already turned in for the night. Harold Cottam, the *Carpathia*'s operator, rushed the message to First Officer Dean on the bridge. They both raced down the ladder, through the chart room, and burst into the captain's cabin. Rostron – a stickler for discipline even when half asleep – wondered what the ship was coming to, with people dashing in in this way. They were meant to knock. But before he could reprimand them, Dean blurted the news.

Rostron bolted out of bed, ordered the ship turned, and then – after the order was given – double-checked Cottam:

'Are you sure it is the *Titanic* and requires immediate assistance?'

'Yes, sir.'

'You are absolutely certain?'

'Quite certain.'

'All right, tell him we are coming along as fast as we can.'

Rostron then rushed into the chart room and worked out the *Carpathia*'s new course. As he figured and scribbled, he saw the boatswain's mate pass by, leading a party to scrub down the decks. Rostron told him to forget the decks and prepare the boats for lowering. The mate gaped. Rostron reassured him: 'It's all right; we're going to another vessel in distress.'

In a few moments the new course was set – north 52 west. The *Carpathia* was fifty-eight miles away. At 14 knots she would take four hours to get there. Too long.

Rostron sent for chief engineer Johnstone, told him to

pour it on – call out the off-duty watch . . . cut off the heat and hot water . . . pile every ounce of steam into the boilers.

Next, Rostron sent for First Officer Dean. He told him to knock off all routine work, organize the ship for rescue operations. Specifically, prepare and swing out all boats . . . . rig electric clusters along the ship's side . . . open all gangway doors . . . hook block and line rope in each gangway . . . rig chair slings for the sick and injured, canvas and bags for hauling up children at every gangway . . . drop pilot ladders and side ladders at gangways and along the sides . . . rig cargo nets to help people up . . . prepare forward derricks (with steam in the winches) to hoist mail and luggage aboard . . . and have oil handy to pour down the lavatories on both sides of the ship, in case the sea grew rough.

Then he called the ship's surgeon, Dr McGhee: collect all the restoratives and stimulants on the ship . . . set up first-aid stations in each dining-saloon . . . put the Hungarian doctor in charge of third class . . . the Italian doctor in second . . . McGhee himself in first.

Now it was Purser Brown's turn: see that the chief steward, the assistant purser and himself each covered a different gangway – receive the *Titanic*'s passengers . . . get their names . . . channel them to the proper dining-saloon (depending on class) for medical check.

Finally, another barrage of orders for chief steward Harry Hughes: call out every man . . . prepare coffee for all hands . . . have soup, coffee, tea, brandy and whisky ready for survivors . . . pile blankets at every gangway . . .

convert smoking-room, lounge and library into dormitories for the rescued . . . group all the *Carpathia*'s steerage passengers together, use the space saved for the *Titanic*'s steerage.

As he gave his orders, Rostron urged them all to keep quiet. The job ahead was tough enough without having the *Carpathia*'s passengers underfoot. The longer they slept, the better. As an extra precaution, stewards were stationed in every corridor. They were to tell any prowling passengers that the *Carpathia* wasn't in trouble, urge them to go back to their cabins.

Then he sent an inspector, the master-at-arms and a special detail of stewards to keep the steerage passengers under control. After all, no one knew how they'd react to being shuffled about.

The ship sprang to life. Down in the engine room it seemed as if everyone had found a shovel and was pouring on the coal. The extra watch tumbled out of their bunks and raced to lend a hand. Most didn't even wait to dress. Faster and faster the old ship knifed ahead – 14 . . . 14.5 . . . 15 . . . 16.5 . . . 17 knots. No one dreamed the *Carpathia* could drive so hard.

In the crew's quarters a tug at his blanket woke up steward Robert H. Vaughan. A voice told him to get up and dress. It was pitch black, but Vaughan could hear his room-mates already pulling on their clothes. He asked what was up, and the voice said the *Carpathia* had hit an iceberg.

Vaughan stumbled to the porthole and looked out. The ship was driving ahead, white waves rolling out from her side. Obviously there was nothing wrong with the *Carpathia.* Bewildered, he and his mates continued dressing – all the more confused because someone had swiped their only light bulb and they had to get ready in the dark.

When they reached the deck, an officer put them to work collecting blankets. Then to the first-class dining-saloon . . . now a beehive of men scurrying about, shifting chairs, resetting tables, moving the liquor from the bar to the buffet. Still Vaughan and his mates couldn't imagine the reason. Elsewhere word spread that Captain Rostron wanted 3,000 blankets for 'that many extra people'. But nobody knew why.

At 1.15 they learned. The stewards were all mustered into the main dining-saloon and chief steward Hughes gave a little speech. He told them about the *Titanic* . . . explained their duties . . . paused . . . then delivered his ending : 'Every man to his post and let him do his full duty like a true Englishman. If the situation calls for it, let us add another glorious page to British history.'

Then the stewards went back to work, most of them now shifting blankets from the bedding lockers to the gangways. These were the men Louis Ogden saw when he first looked out of his cabin. Now he decided to try again. He collared Dr McGhee, who was passing by, but the surgeon only told him, 'Please stay in your cabin – captain's orders.'

'Yes, but what is the matter?'

'An accident, but not to our ship. Stay inside.'

Mr Ogden reported back to his wife. For some reason he was sure the *Carpathia* was on fire and the ship was speeding for help. He began dressing, slipped out on deck, found a quartermaster he knew. This time he got a straight answer: 'There has been an accident to the *Titanic*.'

'You'll have to give me something better than that!' said Ogden, almost triumphantly. 'The *Titanic* is on the northern route and we are on the southern.'

'We're going north like hell. Get back in your room.'

Mr Ogden again reported back to Mrs Ogden, who asked, 'Do you believe it?'

'No. Get up and put on your warmest clothes.' There was no doubt in Mr Ogden's mind now: the *Titanic* was unsinkable; so the surgeon must be covering up. His story confirmed their worst fears – the *Carpathia* was in danger. They must escape. Somehow they managed to sneak out on deck.

Others made it too; and they compared notes together, furtive little groups hiding from their own crew. Gradually they realized the *Carpathia* wasn't in danger. But despite rumours about the *Titanic*, nobody was sure why they were on this wild dash through the night. And of course they couldn't ask or they'd be sent below again. So they just stood there, huddling in the shadows, all eyes straining into the darkness, not even knowing what they were looking for.

In fact, nobody on the *Carpathia* now knew what to look for. In the wireless shack over the second-class smoking-room, Harold Cottam could no longer rouse the *Titanic*. But his set was so miserable – the range was only 150 miles at best – that he wasn't sure what had happened. Perhaps the *Titanic* was still sending, but her signals were now too weak to catch.

On the other hand, the news so far had been all bad. At 1.06, Cottam heard her tell the *Olympic*, 'Get your boats ready; going down fast at the head' . . . at 1.10, 'Sinking head down' . . . at 1.35, 'Engine room getting flooded.'

Once the *Titanic* asked Cottam how long he would take to arrive. 'Say about four hours,' instructed Rostron – he didn't yet realize what the *Carpathia* could do.

Then at 1.50 came a final plea, 'Come as quickly as possible, old man; the engine room is filling up to the boilers.' After that, silence.

Now it was after 2.00, and Cottam still hunched tensely over the set. Once Miss Peterson peeked in at him, noticed that, despite the biting cold, Cottam was still in his shirt-sleeves. He had just started to undress when the first CQD arrived, and he hadn't yet got around to putting on his coat again.

Up on the bridge, Rostron was wondering too. He had organized his men, done everything he could think of, and now came the hardest part of all – waiting. Near him stood Second Officer James Bisset, up forward with extra lookouts. All strained for any sign of ice, any sign of the *Titanic*. But so far there was nothing – just the glassy sea, the blazing stars, the sharp, clear, empty horizon.

At 2.35 Dr McGhee climbed the ladder to the bridge, told Rostron that everything was ready below. As he talked, Rostron suddenly saw the glow of a green flare on the horizon, about half a point off the port bow.

'There's his light!' he shouted. 'He must be still afloat!' It certainly looked that way. The flare was clearly a long way off. To see it at all, it must be high out of the water. It was only 2.40, and they were already in sight – perhaps the *Carpathia* would be in time after all.

Then at 2.45 Second Officer Bisset sighted a tiny shaft of light glistening two points off the port bow. It was the first iceberg – revealed by, of all things, the mirrored light of a star.

Then another berg, then another. Twisting and turning, the *Carpathia* now dodged icebergs on all sides, never slackening speed. On they surged, as the men breathlessly watched for the next berg and from time to time spotted more green flares in the distance.

Now that everything was ready, the stewards had a little free time. Robert Vaughan and his mates went to the afterdeck. Like boxers warming up for a fight, they danced about and playfully rough-housed to keep warm. Once a huge iceberg passed close to starboard, and a man cried, 'Hey fellows! Look at the polar bear scratching himself with a chunk of ice!'

A weak joke perhaps, but the men roared with laughter as the *Carpathia* lunged on.

She was firing rockets now. One every fifteen minutes, with Cunard Roman candles in between. Word spread below that they were in sight. In the main dining-saloon the stewards took up their posts. In the engine room, the stokers shovelled harder than ever. At the gangways and boat stations the men stood ready. Everyone was wild with excitement, and the *Carpathia* herself trembled all over. A sailor later remarked, 'The old boat was as excited as any of us.'

But Rostron's heart was sinking. By 3.35 they were

drawing near the *Titanic*'s position, and still no sign of her. He decided the green flare couldn't have been so high after all. It was just the sparkling-clear night that let him see it from so far off. At 3.50 he put the engines on 'stand by' – they were almost at the spot. At 4.00 he stopped the ship – they were there.

Just then another green flare blazed up. It was directly ahead, low in the water. The flickering light showed the outline of a lifeboat perhaps 300 yards away. Rostron started up his engines, began to manoeuvre the *Carpathia* to starboard so as to pick up the lifeboat on his port side, which was leeward. An instant later he spotted a huge iceberg directly ahead and had to swing the other way to keep from hitting it.

The lifeboat was now to windward, and as he edged towards it, a breeze sprang up and the sea grew choppy. A voice from the dark hailed him, 'We have only one seaman and can't work very well.'

'All right,' Rostron shouted back, and he gently nudged the *Carpathia* closer, until the voice called again, 'Stop your engines!'

It was Fourth Officer Boxhall in boat 2. Sitting beside him, Mrs Walter Douglas of Minneapolis was near hysterics. 'The *Titanic*,' she cried, 'has gone down with everyone on board.'

Boxhall told her to 'shut up', and his sharpness cut her off instantly. She quickly pulled herself together and afterwards always agreed the rebuke was justified.

On the *Carpathia* no one heard her anyhow. All eyes were glued on the lifeboat bobbing towards the gangway.

Mrs Ogden noticed the White Star emblem painted on its side, the lifebelts that made everybody look dressed in white. Mrs Crain wondered about the pale, strained faces looking up at the decks. The only sound was a wailing baby somewhere in the boat.

Lines were dropped, and now the boat was fast. A moment's hesitation, then at 4.10 Miss Elizabeth Allen climbed slowly up the swinging ladder and tumbled into the arms of Purser Brown. He asked her where the *Titanic* was, and she replied it had gone down.

Up on the bridge Rostron knew without asking – yet he felt he had to go through with the formalities. He sent for Boxhall, and as the Fourth Officer stood shivering before him, he put it to him: 'The *Titanic* has gone down?'

'Yes' – Boxhall's voice broke as he said it – 'she went down at about 2.30.'

It was half-day now, and the people on deck could make out other lifeboats on all sides. They were scattered over a four-mile area, and in the grey light of dawn they were hard to distinguish from scores of small icebergs that covered the sea. Mixed with the small bergs were three or four towering monsters, 150 to 200 feet high. To the north and west, about five miles away, stretched a flat, unbroken field of ice as far as the eye could see. The floe was studded here and there with other big bergs that rose against the horizon. 'When I saw the ice I had steamed through during the night,' Rostron later told a friend, 'I shuddered and could only think that some other hand than mine was on that helm during the night.'

The sight was so astonishing, so incredible, that those

who had slept through everything until now couldn't grasp it at all. Mrs Wallace Bradford of San Francisco looked out of her porthole and blinked in disbelief – half a mile away loomed a huge, jagged peak like a rock off-shore. It was not white, and she wondered, 'How in the world can we be near a rock when we are four days out from New York in a southerly direction and in mid-ocean?'

Miss Sue Eva Rule of St Louis was equally puzzled. When she first saw one of the lifeboats splashing through the early dawn, it looked like the gondola of an airship, and the huge grey mound behind it looked like a frame. She was sure they were picking up the crew of a fallen dirigible.

Another bewildered passenger hunted up his steward-ess in the corridor. But she stopped him before he said a word. Pointing to some women tottering into the main dining-saloon she sobbed, 'From the *Titanic*. She's at the bottom of the ocean.'

Ten miles away, with the coming of dawn, life was beginning to stir again on the *Californian*. At 4.00 Chief Officer George Frederick Stewart climbed to the bridge and relieved Second Officer Stone.

Stone brought him up to date – told him about the strange ship, the rockets, the way the stranger disappeared. He added that around 3.40 he saw still another rocket, this time directly south and clearly not from the same ship that fired the first eight. Dead tired, Stone dropped down the ladder and turned in – from now on it was Stewart's headache.

At 4.30 Stewart woke up Captain Lord and began to repeat Stone's story.

'Yes, I know,' interrupted the captain, 'he's been telling me.' Lord had never taken off his uniform, so he now went straight to the bridge and began discussing the best way to work out of the ice field and get on to Boston. Stewart broke in and asked if he wasn't going to check on a ship that was now in sight directly to the south. Lord said, 'No, I don't think so; she's not making any signals now.'

Stewart dropped the matter – he didn't mention that Stone, on his way below, said he was sure the ship to the south couldn't be the same one that fired the first eight rockets.

But he must have thought a good deal more about it, because at 5.40 he woke up wireless operator Evans, who recalled his saying, 'There's a ship been firing rockets. Will you see if you can find out if anything is the matter?'

Evans fumbled in the half-light of day, found the head-phones and tuned in.

Two minutes later Stewart rocketed up the steps to the bridge calling, 'There's a ship sunk!' Then he ran back down to the wireless shack . . . back up again . . . then to Captain Lord with the shattering news: 'The *Titanic* has hit a berg and sunk!'

Captain Lord did just what a good skipper should do. He immediately started his engines and headed for the *Titanic*'s last position.

10. 'Go Away – We Have Just Seen Our Husbands Drown'

'Oh, Muddie, look at the beautiful North Pole with no Santa Claus on it,' little Douglas Spedden said to his mother, Mrs Frederick O. Spedden, as boat 3 threaded its way through the loose ice towards the *Carpathia*.

In fact, the world did look like a picture from a child's book about the Arctic. The sun was just edging over the horizon, and the ice sparkled in its first long rays. The bergs looked dazzling white, pink, mauve, deep blue, depending on how the rays hit them and how the shadows fell. The sea was now bright blue, and little chunks of ice, some no bigger than a man's fist, bobbed in the choppy water. Overhead, the eastern sky was gold and blue, promising a lovely day. The shadows of night lingered in the west – Lawrence Beesley remembered watching the Morning Star shine long after the others had faded. Near the horizon a thin, pale crescent moon appeared.

'A new moon! Turn your money over, boys! That is, if you have any!' fireman Fred Barrett shouted cheerfully to the crew rowing No. 13. Whoops and yells of relief erupted from all the boats, as the men tried to outrow each other in reaching the *Carpathia*. Some began singing, 'Pull for the Shore, Boys'. Some gave organized cheers. Some, however, remained silent – stunned by the sinking or overwhelmed by relief.

'It's all right, ladies, do not grieve. We are picked up.' Lookout Hogg sought to encourage the women staring bleakly ahead in No. 7, but they kept very quiet.

There were no cheers on overturned collapsible B either. Lightoller, Gracie, Bride, Thayer and the others were too busy trying to stay afloat. Stirred by the morning breeze, the waves now washed over the hulk and rocked it back and forth. Every time it rolled, a little more air escaped, and the keel sank still lower into the water. With Lightoller shouting directions, the men still shifted their weight back and forth, but after an hour of this they were dead tired.

The sight of the *Carpathia* arriving with the dawn – so thrilling to everyone else – now meant little to these men. She had stopped four miles away, and they wondered how they could last until they were spotted. Suddenly, as the light spread over the sea, they saw new hope. About 800 yards off, boats 4, 10, 12 and D were still strung together in a line, just as Fifth Officer Lowe had ordered.

The men on collapsible B shouted, 'Ship ahoy!' – but they were too far away to be heard. Then Lightoller fished an officer's whistle out of his pocket and blew a shrill blast. The sound not only carried but told the crew manning the boats that an officer was calling.

In No. 12, seaman Frederick Clinch quickly looked up ... thought he saw about twenty men in the distance standing on, of all things, a ship's funnel. In No. 4, trimmer Samuel Hemming looked over too; and in the early morning light it seemed to him some men were standing on a slab of ice. Little matter, the two boats at once cast off and headed over. It was slow rowing, and as they crept within hailing distance, Lightoller urged them on: 'Come over and take us off!'

'Aye, aye, sir,' somebody called back, and finally the two boats arrived. They were barely in time. By now collapsible B was so delicately balanced that the wash from No. 4 almost swept everybody off. It took all Quartermaster Perkis's skill to manoeuvre the boat safely alongside. On B, Lightoller cautioned the men not to scramble. Even so, the boat gave a sickening roll as each man leaned forward to jump.

One by one they made it. Jack Thayer was so preoccupied with getting safely into No. 12 that he didn't notice his mother right alongside in No. 4. And Mrs Thayer was so numbed by cold and misery that she didn't notice her son. When Colonel Gracie's turn came, he crawled hands first into No. 12, preferring pinched fingers to the risk of a jump. Baker Joughin, still treading water, didn't worry at all. He simply let go Maynard's hand and paddled over to No. 4, where they pulled him in, still thoroughly insulated by his whisky.

Lightoller was last to leave the overturned collapsible. When all the others were transferred, he lifted a lifeless body into No. 12, jumped in himself, and took charge of the boat. It was just about 6.30 when he finally shoved off from the empty keel and began rowing towards the *Carpathia*.

Meanwhile Fifth Officer Lowe gave up his search for swimmers among the wreckage. In an hour's hard work No. 14 picked up only four men, and he knew he was too late to find any more. No man could last longer in the ice-cold water. Now day was breaking and rescue was at hand, Lowe decided to head back for the boats he had left tied together and shepherd them in to the *Carpathia*.

'Hoist a sail forward,' he ordered seaman F. O. Evans as the breeze quickened. In every other boat the crew regarded the mast as an extra encumbrance and the sail as just something that got in the way. In some cases they dumped out this equipment before leaving the *Titanic*; in others it stayed in, and the men cursed as they stumbled over useless spars in the dark. They didn't know how to sail anyhow.

Lowe was different. As he later explained, few seamen were boatmen and few boatmen were seamen, but he was both. Years spent windjamming along the Gold Coast now paid off as he skilfully tacked back and forth. The bow slammed down on the waves, and the spray glittered in the early morning sun as No. 14 bowled along at four knots.

By the time he got back, his little fleet had scattered. Boats 4 and 12 were off picking up the men on B, and Nos. 10 and D were heading separately for the *Carpathia*. D looked in bad shape – low in the water and few oars at work. 'Well,' said Lowe to himself, 'I will go down and pick her up and make sure of her.'

'We have about all we want!' Hugh Woolner shouted as No. 14 sailed up. Lowe tossed over a line and gave them a tow.

Then, about a mile and a half away, he spied collapsible A, completely swamped and making no headway at all.

The people in A never did manage to get the sides up, and now the gunwales lay flush with the water. Of some thirty who originally swam to the boat, most had fallen overboard, numb with the cold. Only a dozen men and third-class passenger Mrs Rosa Abbott were left, standing in freezing water up to their knees.

Lowe arrived just in time . . . took them all aboard No. 14 . . . then set sail again for the *Carpathia*, still towing D. Collapsible A was left behind – abandoned and empty, except for the bodies of three men (with lifebelts covering their faces), R. Norris Williams Jr's fur coat and a ring belonging to third-class passenger Edvard P. Lindell of Helsingborg, Sweden, whom no one remembered seeing all night.

One by one the boats crept up to the *Carpathia*. It was 4.45 when No. 13 made fast, and Lawrence Beesley climbed a rope ladder to the C deck companionway. He felt overwhelmed with gratitude, relief and joy to feel a solid deck under his feet again. Close behind climbed Dr Washington Dodge, who remembered to bring along his lifebelt as a memento.

Mrs Dodge and five-year-old Washington Jr arrived at 5.10 in No. 7. The little boy was hauled up in a mail sack and plopped on to the deck. A steward rushed up with coffee, but Master Dodge announced he would rather have cocoa. The steward promptly dashed off and got some – British liners aren't famous for their service for nothing.

Then came No. 3 at 6.00. Mr and Mrs Spedden climbed aboard immaculately dressed. Close behind came the

Henry Sleeper Harpers, dragoman Hamad Hassah, and Pekingese Sun Yat-Sen. Mr Harper soon discovered Mr Ogden on deck, greeting him with classic detachment: 'Louis, how do you keep yourself looking so young?'

Elizabeth Shutes, arriving in the same boat, didn't try the ladder. She sat in a rope sling, felt herself swept aloft with a mighty jerk. From somewhere above a voice called, 'Careful, fellows, she's a lightweight.'

Bruce Ismay stumbled aboard around 6.30, mumbling 'I'm Ismay . . . I'm Ismay.' Trembling, he stood near the gangway, his back against a bulkhead. Dr McGhee gently approached him: 'Will you not go into the saloon and get some soup or something to drink?'

'No, I really don't want anything at all.'

'Do go and get something.'

'If you will leave me alone, I'll be much happier here,' Ismay blurted, then changed his mind: 'If you can get me in some room where I can be quiet, I wish you would.'

'Please,' the doctor softly persisted, 'go to the saloon and get something hot.'

'I would rather not.'

Dr McGhee gave up. He gently led Ismay to his own cabin. During the rest of thc trip Ismay never left the room; he never ate anything solid; he never received a visitor (except Jack Thayer, once); he was kept to the end under the influence of opiates. It was the start of a self-imposed exile from active life. Within a year he retired from the White Star Line, purchased a large estate on the west coast of Ireland and remained a virtual recluse till he died in 1937.

Olaus Abelseth reached the deck about 7.00. A hot blanket was thrown over his soaked, shivering shoulders and he was rushed to the dining-saloon for brandy and coffee. Mrs Charlotte Collyer and the others in No. 14 tagged along, while Fifth Officer Lowe remained behind, unshipping the mast and stowing the sail. He liked a tidy boat.

And so they came, one boatload after another. As each drew alongside, the survivors already aboard peered down from the promenade deck, searching for familiar faces. Billy Carter stood next to the Ogdens, frantically watching for his wife and children. When the rest of the family finally came alongside in No. 4, Mr Carter leaned far over the rail: 'Where's my son? Where's my son?'

A small boy in the boat lifted a girl's big hat and called 'Here I am, Father.' Legend has it that John Jacob Astor himself placed the hat on the ten-year-old's head, saying in answer to objections, 'Now he's a girl and he can go.'

Washington Dodge was another man who had an agonizing wait for his family – thanks largely to a mischievous streak in five-year-old Washington Jr. Dr Dodge didn't see his wife and son come aboard – nor did Mrs Dodge see her husband on deck, but young Washington did. And he decided it would be great fun to keep it to himself. So he didn't tell his mother and effectively hid from his father. Finally, the Dodges' ever-faithful dining-saloon steward Ray spoiled everything by bringing about a reunion.

The crowds along the rail grew steadily as the *Carpathia*'s own passengers poured from their cabins. Some of them learned in curious ways. Mr and Mrs Charles

Marshall were awakened by the steward knocking on their stateroom door.

'What is it?' called Mr Marshall.

'Your niece wants to see you, sir,' came the answer.

Mr Marshall was nonplussed. All three of his nieces were, he knew, making the *Titanic*'s maiden voyage. They even sent him a wireless last night. How could one of them be on board the *Carpathia*? The steward explained. Minutes later the Marshalls were holding a family reunion with Mrs E. D. Appleton (the other nieces arrived later), and their daughter Evelyn dashed on deck to see the sight.

A strange sight it was. The endless plain of packed ice to the north and west – the big bergs and smaller growlers that floated like scouts in advance of the main floe – gave the sea a curiously busy look. The boats that rowed in from all directions seemed incredibly out of place here in mid-Atlantic.

And the people that straggled from them couldn't have looked more peculiar – Miss Sue Eva Rule noticed one woman wearing only a Turkish towel around her waist and a magnificent fur evening cape over her shoulders. The costumes were a rag-bag of lace-trimmed evening dresses . . . kimonos . . . fur coats . . . plain woollen shawls . . . pyjamas . . . rubber boots . . . white satin slippers. But it was still an age of formality – a surprising number of the women wore hats and the men snap-brim tweed caps.

Strangest of all was the silence. Hardly a word was spoken. Everyone noticed it; everyone had a different explanation. The Reverend P. M. A. Hoques, a passenger on the *Carpathia*, thought people were too horror-stricken to speak. Captain Rostron thought everybody was just too busy. Lawrence Beesley felt they were neither too stunned nor too busy – they were simply in the presence of something too big to grasp.

Occasionally there was a minor commotion. Miss Peterson noticed a little girl named Emily sitting on the promenade deck, sobbing, 'Oh, Mama, Mama, I'm sick. Oh, Mama, Mama!'

While No. 3 was unloading its passengers, a woman clad only in a nightgown and kimono suddenly sat up in the bottom of the boat. Pointing at another lady being hoisted up in a boatswain's chair, she cried: 'Look at that horrible woman! Horrible! She stepped on my stomach. Horrible creature!'

And in the third-class dining-saloon an Italian woman went completely to pieces – sobbing, screaming, banging her fists on the table. Over and over and over she cried, '*Bambino!*' An Italian steward coaxed out the information that both her babies were missing. One was soon located, but she held up two fingers and the hysterics started again. Finally the other was found too – in the pantry on the hot press, where it had been left to thaw out.

By 8.15 all the boats were in except No. 12. It barely moved, still several hundred yards away. The breeze grew stiff, and the sea grew rougher. The crowded gunwales were almost level with the waves – nearly seventy-five people were jammed in. The crowd at the *Carpathia*'s rail watched breathlessly as Lightoller nursed it along.

He was nearly frozen – his uniform soaked and stiff. Around his shoulders he wore a cape with a monk's hood, donated by Mrs Elizabeth Mellenger. At his feet, Mrs Mellenger's thirteen-year-old daughter Madeleine gazed up in admiration. She has cherished that cape and hood ever since.

The people in the boat huddled tightly together . . . trying to keep dry, praying they might make it. At a time like this a man notices little trivial things. As Colonel Gracie worked in vain to revive a lifeless body lying beside him, he wondered why the person wore long, grey woollen stockings.

Now 8.20, and they were only 200 yards off. Rostron, trying to help, turned the *Carpathia*'s bow to within 100 yards. As Lightoller struggled to cross the bow and get in her lee, a sudden squall whipped up the sea. First one wave, then another crashed into the boat. A third just missed. Next instant he was there – safe in the shelter of the big ship.

At 8.30, No. 12 – the last boat to arrive – made fast and began to unload. Colonel Gracie felt like falling down on his knees and kissing the deck as he stepped into the gangway. Harold Bride felt a pair of strong hands reach out to him; then he passed out. Jack Thayer saw his mother waiting and rushed to her arms. Mrs Thayer stammered, 'Where's Daddy?'

'I don't know, Mother,' he answered quietly.

Meanwhile Rostron wondered where to take his 705 unexpected guests. Halifax was nearest, but there was ice along the way, and he thought the *Titanic*'s passengers had seen enough. The Azores were best for the *Carpathia*'s schedule, but he didn't have the linen and provisions to last that far. New York was best for the survivors but most costly to the Cunard Line. He dropped down to the surgeon's cabin where Dr McGhee was examining Bruce Ismay. The man was shattered – anything Rostron wanted was all right with him. So Rostron decided on New York.

Then the *Olympic* broke in: Why not transfer the *Titanic*'s survivors to her? Rostron thought this was an appalling idea – he couldn't see subjecting these people to another transfer at sea. Besides, the *Olympic* was the *Titanic*'s sister ship and the sight alone would be like a hideous ghost. To be on the safe side, he trotted back to Dr McGhee's cabin, checked again with Ismay. The White Star president shuddered at the thought.

So New York it was, and the sooner the better. By now the *Californian* was standing by, and Rostron arranged for her to search the scene while he made for port with the survivors. Then he hauled aboard as many of the *Titanic*'s lifeboats as possible – six on the forward deck, seven in the *Carpathia*'s own davits. The rest were set adrift.

Before heading back, Rostron couldn't resist one last look around. He was a thorough man; he didn't want to overlook the smallest chance. Let the *Californian* go through the motions, but if there was any real hope of picking anybody else up, Rostron wanted the *Carpathia* to do it.

As he cruised, it occurred to him that a brief service might be appropriate. He dropped down and asked if Ismay had any objection. It was always the same – anything Rostron wanted to do was all right with him.

So Rostron sent for the Reverend Father Anderson, an

Episcopal clergyman aboard, and the people from the *Titanic* and *Carpathia* assembled together in the main lounge. There they gave thanks for the living and paid their respects to the lost.

While they murmured their prayers, the *Carpathia* steamed slowly over the *Titanic*'s grave. There were few traces of the great ship – patches of reddish-yellow cork ... some steamer chairs ... several white pilasters ... cushions ... rugs ... lifebelts ... the abandoned boats ... just one body.

At 8.50 Rostron was satisfied. There couldn't possibly be another human being alive. He rang 'full speed ahead' and turned his ship for New York.

Already the city was wildly excited. When the first word arrived at 1.20 a.m., nobody knew what to think. The AP flash was certainly cryptic – just a message from Cape Race that at 10.25 local time the *Titanic* called CQD, reported striking an iceberg, and asked for help immediately. Then another message that the liner was down at the head and putting the women off in boats. Then silence.

The news was in time for the first morning editions – but barely. No leeway for double-checking: only time to decide how to handle it. The story seemed fantastic. Yet there it was. The editors nibbled gingerly; the *Herald*'s headline was typical:

THE NEW TITANIC STRIKES ICE
AND CALLS FOR AID.
VESSELS RUSH TO HER SIDE

Only *The Times* went out on a limb. The long silence after the first few messages convinced managing editor Carr Van Anda that she was gone. So he took a flyer – early editions reported the *Titanic* sinking and the women off in lifeboats; the last edition said she had sunk.

By 8.00 a.m. newsmen were storming the White Star Line offices at 9 Broadway. Vice-President Philip A. S. Franklin made light of the reports: even if the *Titanic* had hit ice, she could float indefinitely. 'We place absolute confidence in the *Titanic*. We believe that the boat is unsinkable.'

But at the same time he was frantically wiring Captain Smith: 'Anxiously await information and probable disposition of passengers.'

By mid-morning friends and relatives of the *Titanic*'s passengers were pouring in: Mrs Benjamin Guggenheim and her brother De Witt Seligman . . . Mrs Astor's father W. H. Force . . . J. P. Morgan Jr . . . hundreds of people nobody recognized. Rich and poor, they all got the same reassuring smiles – no need to worry . . . the *Titanic* was unsinkable; well, anyhow, she could float two or three days . . . certainly there were enough boats for everybody.

And the Press joined in. The *Evening Sun* ran a banner headline:

ALL SAVED FROM TITANIC AFTER COLLISION

The story reported all passengers transferred to the *Parisian* and the *Carpathia*, with the *Titanic* being towed by the *Virginian* to Halifax.

Even business seemed confident. At the first news the reinsurance rate on the *Titanic*'s cargo soared to fifty per cent, then to sixty per cent. But as optimism grew, London rates dropped back to fifty per cent, then to forty-five . . . thirty . . . and finally twenty-five per cent.

Meanwhile Marconi stock skyrocketed. In two days it soared fifty-five points to 225. Not bad for a stock that brought only two dollars just a year ago. And IMM – the great combine that controlled the White Star Line – was now recovering after a shaky start in the morning.

Yet rumours were beginning to spread. No official word, but wireless men listening in on the Atlantic traffic picked up disturbing messages not meant for their ears, and relayed the contents anyhow. During the afternoon a Cunard official heard from a friend downtown that the *Titanic* was definitely gone. A New York businessman wired a friend in Montreal the same thing. Franklin heard too, but the source seemed unreliable; so he decided to keep quiet.

At 6.15 the roof fell in. Word finally arrived from the *Olympic*: the *Titanic* went down at 2.20 a.m.; the *Carpathia* picked up all the boats and was returning to New York with the 675 survivors. The message had been delayed in transit several hours. Nobody knows why, but there has never been any evidence supporting the *World*'s suggestion that it was the work of Wall Street bears and shippers reinsuring their cargoes.

Franklin was still steeling himself to tell the public when the clock in the White Star office struck seven. An alert reporter smelled the gloom in the air, took a chance, and barged into the manager's office. Others followed.

'Gentlemen,' Mr Franklin stammered, 'I regret to say that the *Titanic* sank at 2.20 this morning.'

At first that was all he would say, but bit by bit the reporters chipped out admissions. At 8.00: the *Olympic*'s message 'neglected to say that all the crew had been saved'. At 8.15: 'Probably a number of lives had been lost.' At 8.45: 'We very much fear there has been a great loss of life.' By 9.00 he couldn't keep up the front any longer: it was a 'horrible loss of life' . . . they could replace the ship but 'never the human lives'.

At 10.30 Vincent Astor arrived and disappeared into Franklin's office. In a little while he left weeping. On a hunch a reporter phoned Mrs John Jacob Astor's father, W. H. Force. 'Oh, my God,' cried the old gentleman, 'don't tell me that! Where did you get that report from? It isn't true! It can't be true!'

No one could reach the Strauses' daughter, Mrs Alfred Hess. Early that afternoon she had taken the special train chartered by the White Star Line to meet the supposedly crippled *Titanic* at Halifax. By 8.00 the train was lumbering through the Maine countryside, as Mrs Hess sat in the diner chatting with reporters. She was the only woman on board, and it was rather fun.

She was just starting some grapefruit when the train slowed, stopped, and then began moving backwards. It never stopped until Boston. There she learned, 'Plans have changed; the *Titanic*'s people are going straight to New York.' So she took the sleeper back and was met at the gate by her brother early the next morning: 'Things look pretty bad.'

By now the first survivor list was up, and crowds again stormed the White Star office. Mrs Frank Farquharson and Mrs W. H. Marvin came together to learn about their children, who were coming back from their honeymoon. The bride's mother, Mrs Farquharson, gave a happy little yelp when she spied the name 'Mrs Daniel Marvin'; then managed to stifle it when she saw no 'Mr' beside it.

Mrs Ben Guggenheim clung to the hope that some lifeboat was missing. 'He may be drifting about!' she sobbed.

And he might have been, for all anyone knew. Nobody could get any information out of the *Carpathia* – Rostron was saving his wireless for official traffic and private messages from the survivors – so the newspapers made up their stories. The *Evening World* told of a fog, the *Titanic*'s booming siren, a crash like an earthquake. The *Herald* described how the ship was torn asunder, plunged into darkness, almost capsized at the moment of impact.

When imagination ran low, the papers took it out on the silent rescue ship. The *Evening Mail* thundered:

WATCHERS ANGERED BY CARPATHIA'S SILENCE

The *World* pouted:

CARPATHIA LETS NO SECRETS OF THE TITANIC'S LOSS ESCAPE BY WIRELESS

So Tuesday turned to Wednesday . . . and Wednesday to Thursday . . . and still there was no news. The weeklies were caught now. *Harper's Weekly* described the prominent people aboard, featuring Henry Sleeper Harper, a member of the family who owned the magazine. It conjured a fog and a frightful shock; then remarked a little lamely, 'As to what happened, all is still surmise.' But *Harper's* assured its readers that the rule was women and children first, 'the order long enforced among all decent men who use the sea'. Next issue, the magazine turned a possible embarrassment into a journalistic scoop when Henry Sleeper Harper turned up complete with Pekingese and personal Egyptian dragoman. *Harper's* happily announced an exclusive interview.

Thursday night the wait ended. As the *Carpathia* steamed by the Statue of Liberty, 10,000 people watched from the Battery. As she edged towards pier 54, 30,000 more stood in the waterfront rain. To the end Rostron had no truck with newsmen. He wouldn't let them on the ship at quarantine, and as the *Carpathia* steamed up the North River, tugs chugged beside her, full of reporters shouting questions through megaphones.

At 8.37 she reached the pier and began unloading the *Titanic*'s lifeboats so she could be warped in. They were rowed off to the White Star pier, where souvenir hunters picked them clean during the night. (The next day men were put to work removing the *Titanic*'s nameplate from each boat.)

At 9.35 the *Carpathia* was moored, the gangplank lowered, and the first survivors tumbled off. Later a brown canvas carry-all, its two-by-three-inch sides bulging, was taken off and placed under Customs letter G. Customs officials said it was the only luggage saved from the *Titanic*. Owner Samuel Goldenberg denied such foresight. He claimed he bought it on board the *Carpathia*. He said it contained only the clothes he wore off the *Titanic* and a few accessories purchased on the rescue ship – pyjamas, coat, trousers, dressing gown, raincoat, slippers, two rugs, shirt, collars, toilet goods, and shoes for his wife and himself.

The *Carpathia*'s arrival made clear who survived, but it didn't unravel what had happened. The survivors added their own myths and fables to the fiction conjured up on shore. For some the heartbreaking trip back was too much. Others were simply carried away by the excitement. The more expansive found themselves making a good story even better. The more laconic had their experiences improved by reporters. Some were too shocked, some too ashamed.

Newspaper interviews reported that second-class passenger Emilio Portaluppi rode a cake of ice for hours . . . Miss Marie Young saw the iceberg an hour before the collision . . . seamen Jack Williams and William French watched six men shot down like dogs . . . Philadelphia banker Robert W. Daniel took over the *Carpathia*'s wireless during the trip back. All the evidence went against such stories, but the public was too excited to care.

The sky was the limit. The 19 April New York *Sun* had first-class passenger George Brayton saying:

'The moon was shining and a number of us who were

enjoying the crisp air were promenading about the deck. Captain Smith was on the bridge when the first cry from the lookout came that there was an iceberg ahead. It may have been 300 feet high when I saw it. It was probably 200 yards away and dead ahead. Captain Smith shouted some orders ... a number of us promenaders rushed to the bow of the ship. When we saw we could not fail to hit it, we rushed to the stern. Then came a crash, and the passengers were panic-stricken ... The accident happened at about 10.30 p.m ... about midnight, I think, came the first boiler explosion. Then for the first time, I think, Captain Smith began to get worried ...'

Carpathia seaman Jonas Briggs's interview told the story of Rigel, a handsome black Newfoundland dog, who jumped from the deck of the sinking *Titanic* and escorted a lifeboat to the *Carpathia*, his joyous barks signalling Captain Rostron that he was coming.

Personal thoughts weighed heavily on the minds of some. Lookout Reginald Lee – it seemed a century since that dreadful moment when his mate Fleet sighted the berg – told of a haze on the horizon, remembered Fleet saying, 'Well, if we can see through that, we'll be lucky.' Fleet never recalled the conversation.

An interview with one of the men in first class gave this careful explanation of his presence in No. 7, the first boat to leave:

'On one point all the women were firm. They would not enter a lifeboat until all the men were in first. They feared to trust themselves to the seas in them. It required courage to step into the frail craft as they swung from the creaking davits. Few men were willing to take the chance. An officer rushed behind me and shouted, "You're big enough to pull an oar. Jump into this boat or we'll never get the women off." I was forced to do so, though I admit the ship looked a great deal safer to me than any small boat.'

Gradually the full story emerged, but many of the engaging tales born these first few days have lingered ever since – the lady who refused to leave her Great Dane . . . the band playing 'Nearer My God to Thee' . . . Captain Smith and First Officer Murdoch committing suicide . . . Mrs Brown running No. 6 with a revolver.

But legends are part of great events, and if they help keep alive the memory of gallant self-sacrifice, they serve their purpose. At the time, however, no legends were needed to drive home the story. People were overwhelmed by the tragedy. Flags everywhere flew half-mast. Macy's and the Harris theatres were closed. The French line called off a reception on the new SS *France*. In Southampton, where so many of the crew lived, grief was staggering – twenty families in one street bereaved. Montreal called off a military review. King George and President Taft exchanged condolences – and the Kaiser got into the act. J. S. Bache & Co. cancelled its annual dinner. J. P. Morgan called off the inauguration of a new sanatorium he was building at Aix-les-Bains.

Even the Social Register was shaken. In those days the ship that people travelled on was an important yardstick in measuring their standing, and the Register dutifully kept track. The tragedy posed an unexpected problem. To say that listed families crossed on the *Titanic* gave them their social due, but it wasn't true. To say they arrived on the plodding *Carpathia* was true, but socially misleading. How to handle this dilemma? In the case of those lost, the Register dodged the problem – after their names it simply noted the words, 'died at sea, 15 April 1912'. In the case of the living, the Register carefully ran the phrase, 'Arrived *Titan-Carpath*, 18 April 1912'. The hyphen represented history's greatest sea disaster.

What troubled people especially was not just the tragedy – or even its needlessness – but the element of fate in it all. If the *Titanic* had heeded any of the six ice messages on Sunday . . . if ice conditions had been normal . . . if the night had been rough or moonlit . . . if she had seen the berg fifteen seconds sooner – or fifteen seconds later . . . if she had hit the ice any other way . . . if her watertight bulkheads had been one deck higher . . . if she had carried enough boats . . . if the *Californian* had only come. Had any one of these 'if's turned out right, every life might have been saved. But they all went against her – a classic Greek tragedy.

These thoughts were yet to come, as the *Carpathia* turned towards New York in the bright morning sunshine of 15 April. At this point the survivors still slumped exhausted in deck chairs or sipped coffee in the dining-saloon or absently wondered what they would wear.

The *Carpathia*'s passengers pitched in gallantly – digging out extra toothbrushes, lending clothes, sewing smocks for the children out of steamer blankets brought along in the lifeboats. A Macy's wine buyer bound for

Portugal became a sort of guardian angel for the three rescued Gimbels buyers. Mrs Louis Ogden took cups of coffee to two women in gay coats and scarves sitting alone in a corner. 'Go away,' they said, 'we have just seen our husbands drown.'

For some of the survivors life began again – Lawrence Beesley busily scribbled off a wireless message that he was safe. For others it took longer. Colonel Gracie lay under a pile of blankets on a sofa in the dining-saloon while his clothes dried in the bake oven. Bruce Ismay sat trembling in the surgeon's cabin, shot full of opiates. Harold Bride came to lying in somebody's stateroom; a woman was bending over him, and he felt her hand brushing back his hair and rubbing his face.

Jack Thayer was in another cabin nearby. A kindly man had lent him pyjamas and a bunk. Now Thayer was getting into bed, just as he had started to do ten hours before. He climbed between the cool sheets, and it occurred to him that a cup of brandy he just swallowed was his first drink of hard liquor. He must indeed be growing up.

Far below, the *Carpathia*'s engines hummed with a swift, soothing rhythm. Far above, the wind whistled through the rigging. Ahead lay New York, and home in Philadelphia. Behind, the sun caught the bright red-and-white stripes of the pole from the *Titanic*'s barber shop, as it bobbed in the empty sea. But Jack Thayer no longer knew or cared. The brandy had done its work. He was fast asleep.

Facts about the *Titanic*

'There will never be another like her,' says baker Charles Burgess, who ought to know. In forty-three years on the Atlantic run he has seen them all – *Olympic* . . . *Majestic* . . . *Mauretania* . . . and so on. Today, as carver in the kitchen of the *Queen Elizabeth*, Burgess is probably the last *Titanic* crewman on active service.

'Like the *Olympic*, yes, but so much more elaborate,' he reflects. 'Take the dining-saloon. The *Olympic* didn't even have a carpet, but the *Titanic* – ah, you sank in it up to your knees. Then there's the furniture: so heavy you could hardly lift it. And that panelling . . .

'They can make them bigger and faster, but it was the care and effort that went into her. She was a beautiful, wonderful ship.'

Burgess's reflections are typical. The *Titanic* has cast a spell on all who built and sailed her. So much so that, as the years go by, she grows ever more fabulous. Many survivors now insist she was 'twice as big as the *Olympic*' – actually they were sister ships, with the *Titanic* just 1,004 tons larger. Others recall golf courses, regulation tennis courts, a herd of dairy cows and other little touches that exceeded even the White Star Line's penchant for luxury.

The *Titanic* was impressive enough without embellishment. Her weight – 46,328 gross tons . . . 66,000 tons displacement. Her dimensions – 882.5 feet long . . . 92.5 feet wide . . . 60.5 feet from waterline to boat deck, or 175 feet from keel to the top of her four huge funnels. She was, in short, eleven storeys high and a sixth of a mile long.

Triple screw, the *Titanic* had two sets of four-cylinder reciprocating engines, each driving a wing propeller, and a turbine driving the centre propeller. This combination gave her 50,000 registered horsepower, but she could easily develop at least 55,000 horsepower. At full speed she could make 24 to 25 knots.

Perhaps her most arresting feature was her watertight construction. She had a double bottom and was divided into sixteen watertight compartments. These were formed by fifteen watertight bulkheads running clear across the ship. Curiously, they didn't extend very far up. The first two and the last five went only as high as D deck, while the middle eight were carried only up to E deck. Nevertheless, she could float with any two compartments flooded, and since no one could imagine anything worse than a collision at the conjuncture of two compartments, she was labelled 'unsinkable'.

The 'unsinkable' *Titanic* was launched at the Belfast shipyards of Harland & Wolff on 31 May 1911. The next ten months were spent in fitting her out. She completed her trials on 2 April 1912, and arrived in Southampton on 3 April. A week later she sailed for New York. Here is a reconstructed log of the main events of her maiden voyage:

10 April 1912

12 noon	Leaves Southampton dock; narrowly escapes collision with American liner *New York*.
7.00 p.m.	Stops at Cherbourg for passengers.
9.00 p.m.	Leaves Cherbourg for Queenstown.

11 April 1912

12.30 p.m.	Stops at Queenstown for passengers and mail. One crewman deserts.
2.00 p.m.	Leaves Queenstown for New York, carrying 1,316 passengers and 891 crew.

14 April 1912

9.00 a.m.	*Caronia* reports ice Latitude 42° N from Longitude 49° to 51° W.
1.42 p.m.	*Baltic* reports ice Latitude 41° 51′ N, Longitude 40° 52′ W.
1.45 p.m.	*Amerika* reports ice Latitude 41° 27′ N, Longitude 50° 8′ W.
7.00 p.m.	Temperature 43°.
7.30 p.m.	Temperature 39°.
7.30 p.m.	*Californian* reports ice Latitude 42° 3′ N, Longitude 49° 9′ W.
9.00 p.m.	Temperature 33°.
9.30 p.m.	Second Officer Lightoller warns carpenter and engine room to watch fresh water supply – may freeze up; warns crow's-nest to watch for ice.
9.40 p.m.	*Mesaba* reports ice Latitude 42° N to 41° 25′ N, Longitude 49° to 50° 30′ W.

10.00 p.m.	Temperature 32°.
10.30 p.m.	Temperature of sea down to 31°.
11.00 p.m.	*Californian* warns of ice, but cut off before she gives location.
11.40 p.m.	Collides with iceberg Latitude 41° 46′ N, Longitude 50° 14′ W.

15 April 1912

12.05 a.m.	Orders are given to uncover the boats, muster the crew and passengers.
12.15 a.m.	First wireless call for help.
12.45 a.m.	First rocket fired. First boat, No. 7, lowered.
1.40 a.m.	Last rocket fired.
2.05 a.m.	Last boat, collapsible D, lowered.
2.10 a.m.	Last wireless signals sent.
2.18 a.m.	Lights fail.
2.20 a.m.	Ship founders.
3.30 a.m.	*Carpathia*'s rockets sighted by boats.
4.10 a.m.	First boat, No. 2, picked up by *Carpathia*.
8.30 a.m.	Last boat, No. 12, picked up.
8.50 a.m.	*Carpathia* heads for New York with 705 survivors.

So much for the basic facts. Beyond these, much is a mystery. Probably nothing will ever equal the *Titanic* for the number of unanswered questions she left behind. For instance:

How many lives were lost? Some sources say 1,635 . . . the

American Inquiry, 1,517 . . . the British Board of Trade, 1,503 . . . the British Inquiry, 1,490. The British Board of Trade figure seems most convincing, less fireman J. Coffy, who deserted at Queenstown.

How did various people leave the ship? Nearly every woman survivor who was asked replied firmly, 'in the last boat'. Obviously, all these women didn't go in the same boat, yet to question thc point is like questioning a lady's age – one simply doesn't do it. Careful sifting of the testimony at the British and American hearings shows pretty clearly how the ship was abandoned, but even here there's conflicting evidence. At the British Inquiry each witness was asked how many people were lowered in his lifeboat. The *minimum* estimates were then added. The results show a good deal of wishful thinking:

	Lowered in the boats according to minimum estimates of survivors	*Lowered in the boats according to actual figures of those saved*
Crew	107	139
Men passengers	43	119
Women and children	704	393
Total	854	651

In short, about seventy per cent *more* men and forty-five per cent *fewer* women went in the boats than even the most conservative survivors estimated. Plus the fact that the boats pulled away with twenty-five per cent fewer people than estimated.

What time did various incidents happen? Everyone agrees that the *Titanic* hit the iceberg at 11.40 p.m. and sank at 2.20 a.m. – but there's disagreement on nearly everything that happened in between. The times given in this book are the honest estimates of people intimately involved, but they are far from foolproof. There was simply too much pressure. Mrs Louis M. Ogden, passenger on the *Carpathia*, offers a good example. At one point, while helping some survivors get settled, she paused long enough to ask her husband the time. Mr Ogden's watch had stopped, but he guessed it was 4.30 p.m. Actually, it was only 9.30 in the morning. They were both so engrossed, they had lost all track of time.

What did different people say? There are no reconstructed conversations in this book. The words quoted are given exactly as people remembered them being spoken. Yet there is margin for error. The same conversations are often reported with slight variations. For instance, there are at least four versions of the exchange between Captain Rostron and Fourth Officer Boxhall as boat 2 edged alongside the *Carpathia*. The gist is always the same, but the words vary slightly.

What did the band play? The legend is, of course, that the band went down playing 'Nearer My God to Thee'. Many survivors still insist this was so, and there's no reason to doubt their sincerity. Others maintain the band played only ragtime. One man says he clearly remembers the band in its last moments, and they were not playing at all. In this maze of conflicting evidence, junior wireless operator Harold Bride's story somehow stands out. He was a trained observer, meticulously accurate, and on board to the last. He clearly recalled that, as the boat deck dipped under, the band was playing the Episcopal hymn 'Autumn'.

Did a man get off dressed as a woman? While material was being gathered for this book, four first-class passengers were specifically named as the famous man who escaped in woman's clothes. There is not one shred of evidence that any of these men were guilty, and considerable evidence to the contrary. For instance, investigation suggests that one was the target of a vindictive reporter shoved aside while trying for an interview. Another, prominent in local politics, was the victim of opposition mudslinging. Another was the victim of society gossip; he did happen to leave the *Titanic* before his wife. In the search for bigger game, no one bothered about third-class passenger Daniel Buckley, who freely acknowledged that he wore a woman's shawl over his head. He was only a poor, frightened Irish lad, and nobody was interested.

The answer to all these *Titanic* riddles will never be known for certain. The best that can be done is to weigh the evidence carefully and give an honest opinion. Some will still disagree, and they may be right. It is a rash man indeed who would set himself up as final arbiter on all that happened the incredible night the *Titanic* went down.

Acknowledgements

This book is really about the last night of a small town. The *Titanic* was that big and carried that many people. To tell everything that happened is impossible; to piece even part of the picture together has required the help of literally hundreds of people.

Many of them were there. Some sixty-three survivors were located, and most of these came through handsomely. They are a stimulating mixture of rich and poor, passengers and crew. But all seem to have two qualities in common. First, they look marvellous. It is almost as though, having come through this supreme ordeal, they easily surmounted everything else and are now growing old with calm, tranquil grace. Second, they are wonderfully thoughtful. It seems almost as if, having witnessed man at his most generous, they scorn any trace of selfishness themselves.

Nothing seems to be too much trouble. Many of the survivors have contributed far beyond the scope of the book, just to help me get a better feeling of what it all was like.

For instance, Mrs Noël MacFie (then the Countess of Rothes) tells how – while dining out with friends a year after the disaster – she suddenly experienced the awful feeling of cold and intense horror she always associated with the *Titanic*. For an instant she couldn't imagine why. Then she realized the orchestra was playing 'The Tales of Hoffmann', the last piece of after-dinner music played that fateful Sunday night.

Mrs George Darby, then Elizabeth Nye, similarly contributes an appealing extra touch when she tells how – as it grew bitterly cold early Sunday evening – she and some other second-class passengers gathered in the dining room for a hymn-sing, ending with 'For Those in Peril on the Sea'.

And Mrs Katherine Manning – then Kathy Gilnagh – vividly conveys the carefree spirit of the young people in third class when she talks about the gay party in steerage that same last night. At one point a rat scurried across the room; the boys gave chase; and the girls squealed with excitement. Then the party was on again. Mrs Manning's lovely eyes still glow as she recalls the bagpipes, the laughter, the fun of being a pretty colleen setting out for America.

Most of the survivors, in fact, give glimpses of shipboard life that have an almost haunting quality. You feel it when Mrs G. J. Mercherle (then Mrs Albert Caldwell) recalls the bustle of departing from Southampton . . . when Victorine Perkins (then Chandowson) tells of the Ryersons' sixteen trunks . . . when Mr Spencer Silverthorne remembers his pleasant dinner with the other buyers on Sunday night . . . when Marguerite Schwarzenbach (then Frolicher) describes a quieter supper in her parents' stateroom – she had been sea-sick and this was her first gingerly attempt to eat again.

The crew's recollections have this haunting quality too. You feel it when fireman George Kemish describes the gruff camaraderie of the boiler rooms . . . and when masseuse Maud Slocombe tells of her desperate efforts to get the Turkish bath in apple-pie order. Apparently there was a half-eaten sandwich or empty beer bottle in every nook and corner. 'The builders were Belfast men,' she explains cheerfully.

The atmosphere conveyed by these people somehow contributes as much as the facts and incidents they describe. I appreciate their help enormously.

Other survivors deserve all my thanks for the way they painstakingly reconstructed their thoughts and feelings as the ship was going down. Jack Ryerson searched his mind to recall how he felt as he stood to one side, while his father argued to get him into boat 4. Did he realize his life was hanging in the balance? No, he didn't think very much about it. He was an authentic thirteen-year-old boy.

Washington Dodge Jr's chief impression was the ear-splitting roar of steam escaping from the *Titanic*'s huge funnels. He was an authentic five-year-old.

Third-class passengers Anna Kincaid (then Sjoblom), Celiney Decker (then Yasbeck) and Gus Cohen have also contributed far more than interesting narratives. They have been especially helpful in recreating the atmosphere that prevailed in steerage – a long-neglected side of the story.

The crew too have provided much more than accounts of their experiences. The deep feeling in baker Charles

Burgess's voice, whenever he discusses the *Titanic*, reveals the intense pride of the men who sailed her. The gracious courtesy of stewards James Witter, F. Dent Ray, Alfred Pugh and Leo James Hyland points up the matchless service enjoyed by the passengers. And the thoroughness of men like Quartermaster George Thomas Rowe, baker Walter Belford and greaser Walter Hurst confirms fireman Kemish's boast that the crew were 'the pick of Southampton'.

To these and many other *Titanic* survivors – like Mrs Jacques Futrelle, Mrs Henry B. Harris, Mrs H. A. Cassebeer, Mrs M. V. Mann, Mrs A. C. Williams, Harry Giles, Charles Joughin and Herbert J. Pitman – go my heartfelt thanks for their time and trouble.

The relatives of people on the *Titanic* have been equally cooperative. One letter recently made available by the descendant of a survivor illustrates how far they've gone. It is a letter written to the survivor himself, shortly after the accident. I have left out all names, but the act of making this data available shows a courage and honesty that effectively rebut the charge made in the letter:

Dear –: I have before me information stating that you attempted to force your way into one of the lifeboats . . . and that when ordered back by Major Butt, you slipped from the crowd, disappeared, and after a few moments were seen coming from your stateroom dressed in women's clothing which was recognized as garments worn by your wife en route.

I can't understand how you can hold your head up and call yourself a man among men, knowing that every breath you draw is a lie. If your conscience continues to bother you after reading this, you had better come forward. There is no truer saying than the old one, 'Confession is good for the soul.'

Yours truly, —

Besides making letters available, many of the relatives have supplied fascinating information themselves. Especially, I want to thank Captain Smith's daughter, Mrs M. R. Cooke, for the charming recollection of her gallant father . . . Mrs Sylvia Lightoller for her kindness in writing to me about her late husband, Commander Charles Lightoller, who distinguished himself in 1940 by taking his own boat over to Dunkirk . . . Mrs Alfred Hess for making available the family papers of Mr and Mrs Isidor Straus . . . Mrs Cynthia Fletcher for a copy of the letter written by her father, Hugh Woolner, on board the *Carpathia* . . . Mr Fred G. Crosby and his son John for help in obtaining information on Captain Edward Gifford Crosby . . . and Mrs Victor I. Minahan, for the interesting details about Dr and Mrs William Minahan and their daughter Daisy.

Where survivors and relatives couldn't be found, I have relied on published material. The official transcripts of the Senate's investigation and the British Court of Inquiry of course provide several thousand pages of fascinating testimony. Jack Thayer's privately published reminiscences are an engagingly frank account. Dr Washington Dodge's privately printed speech to the Commonwealth Club in San Francisco is equally interesting. Lawrence Beesley's book, *The Loss of the SS Titanic* (Houghton Mifflin, 1912), contains a classic description worth anybody's time.

Archibald Gracie's *The Truth about the Titanic* (Mitchell Kennerley, 1913) is invaluable for chasing up who went in what boat – Colonel Gracie was an indefatigable detective. Commander Lightoller's *Titanic and other Ships* (Ivor Nicholson & Watson, 1935) mirrors his fine mixture of humour and bravery. Shan Bullock's *A 'Titanic' Hero: Thomas Andrews, Shipbuilder* (Norman-Remington, 1913) is a labour of love, piecing together the last hours of this wonderful man.

Good accounts by some survivors have also appeared from time to time in magazines and newspapers – the anniversary is apparently a godsend to the city desk. Typical – Jack Thayer's story in the Philadelphia *Evening Bulletin* on 14 April 1932 . . . fireman Louis Michelsen's interview in the Cedar Rapids *Gazette* on 15 May 1955 . . . the engaging and lively account by Mrs René Harris in the 23 April 1932 issue of *Liberty* magazine.

The contemporary Press is less satisfactory. *The New York Times* of course did a brilliant job, but most of the New York papers were extremely unreliable. Better work was done by newspapers in towns where local citizens were involved – for instance, the Milwaukee papers on the Crosbys and Minahans . . . the San Francisco papers on the Dodges . . . the Cedar Rapids *Gazette* on the Douglases. Abroad, the London *Times* was thorough if dull. Most fascinating, perhaps, were the papers in Belfast, where the *Titanic* was built, and those in Southampton, where so many of the crew lived. These were seafaring towns, and the coverage had to be good.

The contemporary popular magazines – *Harper's*,

Sphere, *Illustrated London News* – are mostly a rehash of Press stories, but an occasional gem pops up, like Henry Sleeper Harper's description in the 27 April 1912 *Harper's*, or Mrs Charlotte Collyer's fine account in the 26 May 1912 *Semi-Monthly Magazine*. The technical journals of the time offer better pickings – the 1911 special edition of the British magazine *Shipbuilder*, giving complete data on the *Titanic's* construction; and similar material in the 26 May 1911 issue of *Engineering* and the 1 July 1911 issue of *Scientific American*.

The other actors in the drama – the people on the rescue ship *Carpathia* – have been as generous and cooperative as those on the *Titanic*. Mr Robert H. Vaughan has been invaluable in helping to piece together the details of that wild dash through the night. Sir James Bisset and Mr R. Purvis have been especially helpful in recalling the names of various *Carpathia* officers. Mrs Louis M. Ogden has supplied a treasury of anecdotes – all the more valuable because she was one of the first on deck. Mrs Diego Suarez (then Miss Evelyn Marshall) contributed a vivid picture of the scene as the *Titanic's* boats edged alongside.

There is not much published material about the *Carpathia*, but Captain Sir Arthur H. Rostron's book, *Home from the Sea* (Macmillan, 1931), contains an excellent account. His testimony at the United States and British hearings is also valuable, and the same goes for the testimony of wireless operator Harold Thomas Cottam.

In addition to the people on the *Titanic* and *Carpathia*, certain others contributed valuable help in preparing this book. Captain Charles Victor Groves has aided me greatly in piecing together the story of the *Californian*, on which he served as Third Officer. Charles Dienz, who at the time was *maître* of the Ritz Carlton on the *Amerika*, has given valuable information on how these ocean-going *à la carte* restaurants operated – data especially useful since only one of the *Titanic*'s restaurant staff was saved. The Marconi Company has supplied extremely interesting information on wireless installations of the time and Helen Hernandez of Twentieth Century Fox has been a gold mine of useful leads.

Finally, people much closer to me personally deserve a special word of thanks. Mr Ralph Whitney has suggested many useful sources. Mr Harold Daw has contributed important research. Miss Virginia Martin has deciphered and typed reams of scribbled foolscap. My mother has done the kind of painful indexing and cross-checking that only a mother would be willing to do.

W. L., 1956

Passenger List

Here is the White Star Line's final list of lost and saved, dated 9 May 1912. Those saved are in *italics*.

FIRST-CLASS PASSENGERS

Allen, Miss Elizabeth Walton
Allison, Mr H. J.
Allison, Mrs H. J. *and Maid*
Allison, Miss L.
Allison, Master T. and Nurse
Anderson, Mr Harry
Andrews, Miss Cornelia I.
Andrews, Mr Thomas
Appleton, Mrs E. D.
Artagaveytia, Mr Ramon
Astor, Colonel J. J. and Manservant
Astor, Mrs J. J. and Maid
Aubert, Mrs N. and Maid
Barkworth, Mr A. H.
Baumann, Mr J.
Baxter, Mrs James
Baxter, Mr Quigg
Beattie, Mr T.
Beckwith, Mr R. L.
Beckwith, Mrs R. L.
Behr, Mr K. H.
Bishop, Mr D. H.
Bishop, Mrs D. H.
Bjornstrom, Mr H.
Blackwell, Mr Stephen Weart
Blank, Mr Henry
Bonnell, Miss Caroline
Bonnell, Miss Lily
Borebank, Mr J. J.
Bowen, Miss
Bowerman, Miss Elsie
Brady, Mr John B.
Brandeis, Mr E.
Brayton, Mr George
Brewe, Dr Arthur Jackson
Brown, Mrs J. J.
Brown, Mrs J. M.
Bucknell, Mrs W. and Maid
Butt, Major Archibald W.
Calderhead, Mr E. P.
Candee, Mrs Churchill
Cardoza, Mrs J. W. M. and Maid
Cardoza, Mr T. D. M. and Manservant
Carran, Mr F. M.
Carran, Mr J. P.
Carter, Mr William E.

Carter, Mrs William E. and Maid
Carter, Miss Lucile
Carter, Master William T. and
Manservant
Case, Mr Howard B.
Cassebeer, Mrs H. A.
Cavendish, Mr T. W.
Cavendish, Mrs T. W. and Maid
Chaffee, Mr Herbert F.
Chaffee, Mrs Herbert F.
Chambers, Mr N. C.
Chambers, Mrs N. C.
Cherry, Miss Gladys
Chevré, Mr Paul
Chibnall, Mrs E. M. Bowerman
Chisholm, Mr Robert
Clark, Mr Walter M.
Clark, Mrs Walter M.
Clifford, Mr George Quincy
Colley, Mr E. P.
Compton, Mrs A. T.
Compton, Miss S. P.
Compton, Mr A. T., Jr
Cornell, Mrs R. G.
Crafton, Mr John B.
Crosby, Mr Edward G.
Crosby, Mrs Edward G.
Crosby, Miss Harriet
Cummings, Mr John Bradley
Cummings, Mrs John Bradley
Daly, Mr P. D.
Daniel, Mr Robert W.
Davidson, Mr Thornton
Davidson, Mrs Thornton
De Villiers, Mrs B.
Dick, Mr A. A.
Dick, Mrs A. A.
Dodge, Dr Washington
Dodge, Mrs Washington
Dodge, Master Washington
Douglas, Mrs F. C.
Douglas, Mr W. D.
Douglas, Mrs W. D. and Maid
Dulles, Mr William C.
Earnshew, Mrs Boulton
Endres, Miss Caroline
Eustis, Miss E. M.
Evans, Miss E.
Flegenheim, Mrs A.
Flynn, Mr J. I.
Foreman, Mr B. L.
Fortune, Mr Mark
Fortune, Mrs Mark
Fortune, Miss Ethel
Fortune, Miss Alice
Fortune, Miss Mabel
Fortune, Mr Charles
Franklin, Mr T. P.
Frauenthal, Mr T. G.
Frauenthal, Dr Henry W.
Frauenthal, Mrs Henry W.
Frolicher, Miss Marguerite
Futrelle, Mr J.
Futrelle, Mrs J.
Gee, Mr Arthur
Gibson, Mrs L.
Gibson, Miss D.
Giglio, Mr Victor
Goldenberg, Mr S. L.
Goldenberg, Mrs S. L.

Goldschmidt, Mrs George B.
Gordon, Sir Cosmo Duff
Gordon, Lady Duff and Maid
Gracie, Colonel Archibald
Graham, Mr William G.
Graham, Mrs William G.
Graham, Miss Margaret
Greenfield, Mrs L. D.
Greenfield, Mr W. B.
Guggenheim, Mr Benjamin
Harder, Mr George A.
Harder, Mrs George A.
Harper, Mr Henry Sleeper and Manservant
Harper, Mrs Henry Sleeper
Harris, Mr Henry B.
Harris, Mrs Henry B.
Harrison, Mr W. H.
Haven, Mr H.
Hawksford, Mr W. J.
Hays, Mr Charles M.
Hays, Mrs Charles M. and Maid
Hays, Miss Margaret
Head, Mr Christopher
Hilliard, Mr Herbert Henry
Hipkins, Mr W. E.
Hippach, Mrs Ida S.
Hippach, Miss Jean
Hogeboom, Mrs John C.
Holverson, Mr A. O.
Holverson, Mrs A. O.
Hoyt, Mr Frederick M.
Hoyt, Mrs Frederick M.
Hoyt, Mr W. F.
Isham, Miss A. E.
Ismay, Mr J. Bruce and Manservant
Jakob, Mr Birnbaum
Jones, Mr C. C.
Julian, Mr H. F.
Kent, Mr Edward A.
Kenyon, Mr F. R.
Kenyon, Mrs F. R.
Kimball, Mr E. N.
Kimball, Mrs E. N.
Klaber, Mr Herman
Lambert-Williams, Mr Fletcher Fellows
Leader, Mrs F. A.
Lewy, Mr E. G.
Lindstroem, Mrs J.
Lines, Mrs Ernest H.
Lines, Miss Mary C.
Lingrey, Mr Edward
Long, Mr Milton C.
Longley, Miss Gretchen F.
Loring, Mr J. H.
Madill, Miss Georgette Alexandra
Maguire, Mr J. E.
Maréchal, Mr Pierre
Marvin, Mr D. W.
Marvin, Mrs D. W.
McCaffry, Mr T.
McCarthy, Mr Timothy J.
McGough, Mr J. R.
Meyer, Mr Edgar J.
Meyer, Mrs Edgar J.
Millet, Mr Frank D.
Minahan, Dr W. E.
Minahan, Mrs W. E.

Minahan, Miss Daisy
Moch, Mr Philip E.
Molsom, Mr H. Markland
Moore, Mr Clarence and Manservant
Natsch, Mr Charles
Newell, Mr A. W.
Newell, Miss Alice
Newell, Miss Madeline
Newsom, Miss Helen
Nicholson, Mr A. S.
Omont, Mr F.
Ostby, Mr E. C.
Ostby, Miss Helen R.
Ovies, Mr S.
Parr, Mr M. H. W.
Partner, Mr Austin
Payne, Mr V.
Pears, Mr Thomas
Pears, Mrs Thomas
Penasco, Mr Victor
Penasco, Mrs Victor and Maid
Peuchen, Major Arthur
Porter, Mr Walter Chamberlain
Potter, Mrs Thomas, Jr
Reuchlin, Jonkheer J. G.
Rheims, Mr George
Robert, Mrs Edward S. and Maid
Roebling, Mr Washington A., 2nd
Rolmane, Mr C.
Rood, Mr Hugh R.
Rosenbaum, Miss
Ross, Mr J. Hugo
Rothes, the Countess of and Maid
Rothschild, Mr M.
Rothschild, Mrs M.
Rowe, Mr Alfred
Ryerson, Mr Arthur
Ryerson, Mrs Arthur
Ryerson, Miss Emily
Ryerson, Miss Susan
Ryerson, Master Jack
Saalfeld, Mr Adolphe
Schabert, Mrs Paul
Seward, Mr Frederick K.
Shutes, Miss E. W.
Silverthorne, Mr S. V.
Silvey, Mr William B.
Silvey, Mrs William B.
Simonius, Oberst Alfons
Sloper, Mr William T.
Smart, Mr John M.
Smith, Mr J. Clinch
Smith, Mr R. W.
Smith, Mr L. P.
Smith, Mrs L. P.
Snyder, Mr John
Snyder, Mrs John
Soloman, Mr A. L.
Spedden, Mr Frederick O.
Spedden, Mrs Frederick O. and Maid
Spedden, Master R. Douglas and Nurse
Spencer, Mr W. A.
Spencer, Mrs W. A. and Maid
Stahelin, Dr Max
Stead, Mr W. T.
Steffanson, H. B.
Stehli, Mr Max Frolicher

Stehli, Mrs Max Frolicher
Stengel, Mr C. E. H.
Stengel, Mrs C. E. H.
Stephenson, Mrs W. B.
Stewart, Mr A. A.
Stone, Mrs George M. and Maid
Straus, Mr Isidor and Manservant
Straus, Mrs Isidor *and Maid*
Sutton, Mr Frederick
Swift, Mrs Frederick Joel
Taussig, Mr Emil
Taussig, Mrs Emil
Taussig, Miss Ruth
Taylor, Mr E. Z.
Taylor, Mrs E. Z.
Thayer, Mr J. B.
Thayer, Mrs J. B. and Maid
Thayer, Mr J. B., Jr
Thorne, Mr G.
Thorne, Mrs G.
Tucker, Mr G. M., Jr
Uruchurtu, Mr M. R.
Van der Hoef, Mr Wyckoff
Walker, Mr W. Anderson
Warren, Mr F. M.
Warren, Mrs F. M.
Weir, Mr J.
White, Mr Percival W.
White, Mr Richard F. and Manservant
White, Mrs J. Stuart and Maid
Wick, Mr George D.
Wick, Mrs George D.
Wick, Miss Mary
Widener, Mr George D. and Manservant
Widener, Mrs George D. and Maid
Widener, Mr Harry
Willard, Miss Constance
Williams, Mr Duane
Williams, Mr R. N., Jr
Woolner, Mr Hugh
Wright, Mr George
Young, Miss Marie and Maid

SECOND-CLASS PASSENGERS

Abelson, Mr Samson
Abelson, Mrs Hanna
Aldworth, Mr C.
Andrew, Mr Edgar
Andrew, Mr Frank
Angle, Mr William
Angle, Mrs
Ashby, Mr John
Baily, Mr Percy
Baimbrigge, Mr Chas. R.
Balls, Mrs Ada E.
Banfield, Mr Frederick J.
Bateman, Mr Robert J.
Beane, Mr Edward
Beane, Mrs Ethel
Beauchamp, Mr H. J.
Becker, Mrs A. O. and three children
Beesley, Mr Lawrence
Bentham, Miss Lillian W.
Berriman, Mr William
Botsford, Mr W. Hull
Bowenur, Mr Solomon
Bracken, Mr Jas. H.

Brito, Mr Jose de
Brown, Miss Mildred
Brown, Mr S.
Brown, Mrs S.
Brown, Miss E.
Bryhl, Mr Curt
Bryhl, Miss Dagmar
Buss, Miss Kate
Butler, Mr Reginald
Byles, Rev. Thomas R. D.
Bystrom, Miss Karolina
Caldwell, Mr Albert F.
Caldwell, Mrs Sylvia
Caldwell, Master Alden G.
Cameron, Miss Clear
Carbines, Mr William
Carter, Rev. Ernest C.
Carter, Mrs Lillian
Chapman, Mr John H.
Chapman, Mrs Elizabeth
Chapman, Mr Charles
Christy, Mrs Alice
Christy, Miss Juli
Clarke, Mr Charles V.
Clarke, Mrs Ada Maria
Coleridge, Mr R. C.
Collander, Mr Erik
Collett, Mr Stuart
Collyer, Mr Harvey
Collyer, Mrs Charlotte
Collyer, Miss Marjorie
Corbett, Mrs Irene
Corey, Mrs C. P.
Cotterill, Mr Harry
Davis, Mr Charles
Davis, Mrs Agnes
Davis, Master John M.
Davis, Miss Mary
Deacon, Mr Percy
Del Carlo, Mr Sebastian
Del Carlo, Mrs
Denbou, Mr Herbert
Dibden, Mr William
Doling, Mrs Ada
Doling, Miss Elsie
Downton, Mr William J.
Drachstedt, Baron von
Drew, Mr James V.
Drew, Mrs Lulu
Drew, Master Marshall
Duran, Miss Florentina
Duran, Miss Asuncion
Eitemiller, Mr G. F.
Enander, Mr Ingvar
Fahlstrom, Mr Arne J.
Faunthorpe, Mr Harry
Faunthorpe, Mrs Lizzie
Fillbrook, Mr Charles
Fox, Mr Stanley H.
Funk, Miss Annie
Fynney, Mr Jos.
Gale, Mr Harry
Gale, Mr Shadrach
Garside, Miss Ethel
Gaskell, Mr Alfred
Gavey, Mr Lawrence
Gilbert, Mr William
Giles, Mr Edgar
Giles, Mr Fred
Giles, Mr Ralph

Gill, Mr John
Gillespie, Mr William
Givard, Mr Hans K.
Greenberg, Mr Samuel
Hale, Mr Reginald
Hamalainer, Mrs Anna and Infant
Harbeck, Mr Wm H.
Harper, Mr John
Harper, Miss Nina
Harris, Mr George
Harris, Mr Walter
Hart, Mr Benjamin
Hart, Mrs Esther
Hart, Miss Eva
Herman, Miss Alice
Herman, Mrs Jane
Herman, Miss Kate
Herman, Mr Samuel
Hewlett, Mrs Mary D.
Hickman, Mr Leonard
Hickman, Mr Lewis
Hickman, Mr Stanley
Hiltunen, Miss Martha
Hocking, Mr George
Hocking, Mrs Elizabeth
Hocking, Miss Nellie
Hocking, Mr Samuel J.
Hodges, Mr Henry P.
Hoffman, Mr *and two children (Lolo and Louis)*
Hold, Mrs Annie
Hold, Mr Stephen
Hood, Mr Ambrose
Hosono, Mr Masabumi
Howard, Mr Benjamin
Howard, Mrs Ellen T.
Hunt, Mr George
Ilett, Miss Bertha
Jacobson, Mr Sidney S.
Jacobsohn, Mrs Amy F.
Jarvis, Mr John D.
Jefferys, Mr Clifford
Jefferys, Mr Ernest
Jenkin, Mr Stephen
Jervan, Mrs A. T.
Kantor, Mrs Miriam
Kantor, Mr Sehua
Karnes, Mrs J. F.
Keane, Mr Daniel
Keane, Miss Nora A.
Kelly, Mrs F.
Kirkland, Rev. Charles L.
Kvillner, Mr John Henrik
Lahtinen, Mr William
Lahtinen, Mrs Anna
Lamb, Mr J. J.
Lamore, Mrs Amelia
Laroche, Mr Joseph
Laroche, Mrs Juliet
Laroche, Miss Louise
Laroche, Miss Simonne
Lehman, Miss Bertha
Leitch, Miss Jessie
Levy, Mr R. J.
Leyson, Mr Robert W. N.
Lingan, Mr John
Louch, Mr Charles
Louch, Mrs Alice Adela
Mack, Mrs Mary

Malachard, Mr Noel
Mallet, Mr A.
Mallet, Mrs
Mallet, Master A.
Mangiavacchi, Mr Emilio
Mantvila, Mr Joseph
Marshall, Mr
Marshall, Mrs Kate
Matthews, Mr W. J.
Maybery, Mr Frank H.
McCrae, Mr Arthur G.
McCrie, Mr James
McKane, Mr Peter D.
Mellenger, Mrs Elizabeth
Mellenger, Miss M.
Mellers, Mr William
Meyer, Mr August
Milling, Mr Jacob C.
Mitchell, Mr Henry
Morawick, Dr Ernest
Mudd, Mr Thomas C.
Myles, Mr Thomas F.
Nasser, Mr Nicolas
Nasser, Mrs
Nesson, Mr Israel
Nicholls, Mr Joseph C.
Norman, Mr Robert D.
Nye, Mrs Elizabeth
Otter, Mr Richard
Oxenham, Mr P. Thomas
Padro, Mr Julian
Pain, Dr Alfred
Pallas, Mr Emilio
Parker, Mr Clifford R.
Parrish, Mrs L. Davis
Pengelly, Mr Frederick
Pernot, Mr René
Peruschitz, Rev. Jos. M.
Phillips, Mr Robert
Phillips, Miss Alice
Pinsky, Miss Rosa
Ponesel, Mr Martin
Portaluppi, Mr Emilio
Pulbaun, Mr Frank
Quick, Mrs Jane
Quick, Miss Vera W.
Quick, Miss Phyllis
Reeves, Mr David
Renouf, Mr Peter H.
Renouf, Miss Lillie
Reynolds, Miss E.
Richards, Mr Emile
Richards, Mrs Emile
Richards, Master William
Richards, Master George
Ridsdale, Miss Lucy
Rogers, Mr Harry
Rogers, Miss Selina
Rugg, Miss Emily
Sedgwick, Mr C. F. W.
Sharp, Mr Percival
Shelley, Mrs Imanita
Silven, Miss Lyyli
Sincock, Miss Maude
Sinkkenen, Miss Anna
Sjostedt, Mr Ernest A.
Slayter, Miss H. M.
Slemen, Mr Richard J.
Smith, Mr Augustus
Smith, Miss Marion

Sobey, Mr Hayden
Stanton, Mr S. Ward
Stokes, Mr Phillip J.
Swane, Mr George
Sweet, Mr George
Toomey, Miss Ellen
Trant, Miss Jessie
Tronpiansky, Mr Moses A.
Troutt, Miss E. Celia
Turpin, Mrs Dorothy
Turpin, Mr William J.
Veale, Mr James
Walcroft, Miss Nellie
Ware, Mrs Florence L.
Ware, Mr John James
Ware, Mr William J.
Watt, Miss Bertha
Watt, Mrs Bessie
Webber, Miss Susie
Weisz, Mr Leopold
Weisz, Mrs Matilda
Wells, Mrs Addie
Wells, Miss J.
Wells, Master Ralph
West, Mr E. Arthur
West, Mrs Ada
West, Miss Barbara
West, Miss Constance
Wheadon, Mr Edward
Wheeler, Mr Edwin
Wilhelms, Mr Charles
Williams, Mr C.
Wright, Miss Marion
Yrois, Miss H.

THIRD-CLASS PASSENGERS

British subjects embarked at Southampton

Abbing, Anthony
Abbott, Eugene
Abbott, Rosa
Abbott, Rossmore
Adams, J.
Aks, Filly
Aks, Leah
Alexander, William
Allen, William
Allum, Owen G.
Badman, Emily
Barton, David
Beavan, W. T.
Billiard, A. van
Billiard, James (child)
Billiard, Walter (child)
Bing, Lee
Bowen, David
Braund, Lewis
Braund, Owen
Brocklebank, William
Cann, Ernest
Carver, A.
Celotti, Francesco
Chip, Chang
Christmann, Emil
Cohen, Gurshon
Cook, Jacob
Corn, Harry
Coutts, Winnie
Coutts, William (child)

Coutts, Leslie (child)
Coxon, Daniel
Crease, Ernest James
Cribb, John Hatfield
Cribb, Alice
Dahl, Charles
Davies, Evan
Davies, Alfred
Davies, John
Davis, Joseph
Davison, Thomas H.
Davison, Mary
Dean, Mr Bertram F.
Dean, Mrs Hetty
Dean, Bertram (child)
Dean, Vera (infant)
Dennis, Samuel
Dennis, William
Derkings, Edward
Dowdell, Elizabeth
Drapkin, Jenie
Dugemin, Joseph
Elsbury, James
Emanuel, Ethel (child)
Everett, Thomas J.
Foo, Choong
Ford, Arthur
Ford, Margaret
Ford, Miss D. M.
Ford, Mr E. W.
Ford, M. W. T. N.
Ford, Maggie (child)
Franklin, Charles
Garfirth, John
Gilinski, Leslie
Godwin, Frederick
Goldsmith, Emily A.
Goldsmith, Frank J.
Goldsmith, Frank J. W.
Goodwin, Augusta
Goodwin, Lillian A.
Goodwin, Charles E.
Goodwin, William F. (child)
Goodwin, Jessie (child)
Goodwin, Harold (child)
Goodwin, Sidney (child)
Green, George
Guest, Robert
Harknett, Alice
Harmer, Abraham
Hee, Ling
Howard, May
Hyman, Abraham
Johnson, Mr A.
Johnson, Mr W.
Johnston, A. G.
Johnston, Mrs
Johnston, William (child)
Johnston, Miss C. H. (child)
Keefe, Arthur
Kelly, James
Lam, Ali
Lam, Len
Lang, Fang
Leonard, Mr L.
Lester, James
Ling, Lee
Lithman, Simon
Lobb, Cordelia
Lobb, William A.

Lockyer, Edward
Lovell, John
MacKay, George W.
Maisner, Simon
McNamee, Eileen
McNamee, Neal
Meanwell, Marian O.
Meek, Annie L.
Meo, Alfonso
Miles, Frank
Moor, Beile
Moor, Meier
Moore, Leonard C.
Morley, William
Moutal, Rahamin
Murdlin, Joseph
Nancarrow, W. H.
Niklasen, Sander
Nosworthy, Richard C.
Peacock, Alfred (infant)
Peacock, Treasteall
Peacock, Treasteall (child)
Pearce, Ernest
Peduzzi, Joseph
Perkin, John Henry
Peterson, Mairus
Potchett, George
Rath, Sarah
Reed, James George
Reynolds, Harold
Risien, Emma
Risien, Samuel
Robins, Alexander
Robins, Charity
Rogers, William John
Rouse, Richard H.
Rush, Alfred George J.
Sadowitz, Harry
Sage, John
Sage, Annie
Sage, Stella
Sage, George
Sage, Douglas
Sage, Frederick
Sage, Dorothy
Sage, William (child)
Sage, Ada (child)
Sage, Constance (child)
Sage, Thomas (child)
Sather, Simon
Saundercock, W. H.
Sawyer, Frederick
Scrota, Maurice
Shellard, Frederick
Shorney, Charles
Simmons, John
Slocovski, Selman
Somerton, Francis W.
Spector, Woolf
Spinner, Henry
Stanley, Amy
Stanley, Mr E. R.
Storey, Mr T.
Sunderland, Victor
Sutehall, Henry
Theobald, Thomas
Thomson, Alex
Thorneycroft, Florence
Thorneycroft, Percival
Tomlin, Ernest P.

Torber, Ernest
Trembisky, Berk
Tunquist, W.
Ware, Frederick
Warren, Charles W.
Webber, James
Wilkes, Ellen
Willey, Edward
Williams, Harry
Williams, Leslie
Windelov, Einar
Wiseman, Philip

Non-British embarked at Southampton

Abelseth, Karen
Abelseth, Olaus
Abrahamson, August
Adahl, Mauritz
Adolf, Humblin
Ahlin, Johanna
Ahmed, Ali
Alhomaki, Ilmari
Ali, William
Anderson, Alfreda
Anderson, Erna
Anderson, Albert
Anderson, Anders
Anderson, Samuel
Anderson, Sigrid (child)
Anderson, Thor
Anderson, Carla
Anderson, Ingeborg (child)
Anderson, Ebba (child)
Anderson, Sigbard (child)
Anderson, Ellis
Anderson, Ida Augusta
Andreason, Paul Edvin
Angheloff, Minko
Arnold, Joseph
Arnold, Josephine
Aronsson, Ernest Axel A.
Asim, Adola
Asplund, Carl (child)
Asplund, Charles
Asplund, Felix (child)
Asplund, Gustaf (child)
Asplund, Johan
Asplund, Lillian (child)
Asplund, Oscar (child)
Asplund, Selma
Assam, Ali
Augustsan, Albert
Backstrom, Karl
Backstrom, Marie
Balkic, Cerin
Benson, John Viktor
Berglund, Ivar
Berkeland, Hans
Bjorklund, Ernst
Bostandyeff, Guentcho
Braf, Elin Ester
Brobek, Carl R.
Cacic, Grego
Cacic, Luka
Cacic, Maria
Cacic, Manda
Calie, Peter
Carlson, Carl R.
Carlson, Julius

Carlsson, August Sigfrid
Coelho, Domingos Fernardeo
Coleff, Fotio
Coleff, Peyo
Cor, Bartol
Cor, Ivan
Cor, Ludovik
Dahl, Mauritz
Dahlberg, Gerda
Dakic, Branko
Danbom, Ernest
Danbom, Gillber (infant)
Danbom, Sigrid
Danoff, Yoto
Dantchoff, Khristo
Delalic, Regyo
Denkoff, Mito
Dimic, Jovan
Dintcheff, Valtcho
Dyker, Adolf
Dyker, Elizabeth
Ecimovic, Joso
Edwardsson, Gustaf
Eklunz, Hans
Ekstrom, Johan
Finote, Luigi
Fischer, Eberhard
Goldsmith, Nathan
Goncalves, Manoel E.
Gronnestad, Daniel D.
Gustafson, Alfred
Gustafson, Anders
Gustafson, Johan
Gustafsson, Gideon
Haas, Aloisia
Hadman, Oscar
Hagland, Angvald O.
Hagland, Konrad R.
Hakkurainen, Pekko
Hakkurainen, Elin
Hampe, Leon
Hankonen, Eluna
Hansen, Claus
Hansen, Janny
Hansen, Henry Damgavd
Heininen, Wendla
Hendekovic, Ignaz
Henriksson, Jenny
Hervonen, Helga
Hervonen, Hildwe (child)
Hickkinen, Laina
Hillstrom, Hilda
Holm, John F. A.
Holten, Johan
Humblin, Adolf
Ilieff, Ylio
Ilmakangas, Ida
Ilmakangas, Pista
Ivanoff, Konio
Jansen, Carl
Jardin, Jose Netto
Jensen, Carl
Jensen, Hans Peter
Jensen, Svenst L.
Jensen, Nilho R.
Johannessen, Bernt
Johannessen, Elias
Johansen, Nils
Johanson, Oscar
Johanson, Oscar L.

Johansson, Erik
Johansson, Gustaf
Johnson, Jakob A.
Johnson, Alice
Johnson, Harold
Johnson, Eleanor (infant)
Johnsson, Carl
Johnsson, Malkolm
Jonkoff, Lazor
Jonsson, Nielo H.
Jusila, Katrina
Jusila, Mari
Jusila, Erik
Jutel, Henrik Hansen
Kallio, Nikolai
Kalvig, Johannes H.
Karajic, Milan
Karlson, Einar
Karlson, Nils August
Kekic, Tido
Kink, Anton
Kink, Louise
Kink, Louise (child)
Kink, Maria
Kink, Vincenz
Klasen, Klas A.
Klasen, Hilda
Klasen, Gertrud (child)
Laitinen, Sofia
Laleff, Kristo
Landegren, Aurora
Larson, Viktor
Larsson, Bengt Edvin
Larsson, Edvard
Lefebre, Frances
Lefebre, Henry (child)
Lefebre, Ida (child)
Lefebre, Jeannie (child)
Lefebre, Mathilde (child)
Leinonen, Antti
Lindablom, August
Lindahl, Agda
Lindell, Edvard B.
Lindell, Elin
Lindqvist, Einar
Lulic, Nicola
Lundahl, John
Lundin, Olga
Lundstrom, Jan
Madsen, Fridjof
Maenpaa, Matti
Maidenoff, Penko
Makinen, Kalle
Mampe, Leon
Marinko, Dmitri
Markoff, Marin
Melkebuk, Philemon
Messemacker, Guillaume
Messemacker, Emma
Midtsjo, Carl
Mikanen, John
Mineff, Ivan
Minkoff, Lazar
Mirko, Dika
Mitkoff, Mito
Moen, Sigurd H.
Moss, Albert
Mulder, Theo
Myhrman, Oliver
Nankoff, Minko

Nedeco, Petroff
Nenkoff, Christo
Nieminen, Manta
Nilson, Berta
Nilson, Helmina
Nilsson, August F.
Nirva, Isak
Nyoven, Johan
Nyston, Anna
Odahl, Martin
Olsen, Arthur
Olsen, Carl
Olsen, Henry
Olsen, Ole M.
Olson, Elon
Olsson, John
Olsson, Elida
Oreskovic, Luka
Oreskovic, Maria
Oreskovic, Jeko
Orman, Velin
Osman, Mara
Pacruic, Mate
Pacruic, Tome
Panula, Eino
Panula, Ernesti
Panula, Juho
Panula, Maria
Panula, Sanni
Panula, Urhu (child)
Panula, William (infant)
Pasic, Jakob
Paulsson, Alma C.
Paulsson, Gosta (child)
Paulsson, Paul (child)
Paulsson, Stina (child)
Paulsson, Torborg (child)
Pavlovic, Stefo
Pekonemi, E.
Pelsmaker, Alfons de
Peltomaki, Nikolai
Pentcho, Petroff
Person, Ernest
Peterson, Johan
Petersson, Ellen
Petranec, Matilda
Petterson, Olaf
Plotcharsky, Vasil
Radeff, Alexandre
Rintamaki, Matti
Rosblom, Helene
Rosblom, Salli (child)
Rosblom, Viktor
Rummstvedt, Kristian
Salander, Carl
Saljilsvik, Anna
Salonen, Werner
Sandman, Johan
Sandstrom, Agnes
Sandstrom, Beatrice (child)
Sandstrom, Margretha (child)
Sdycoff, Todor
Sheerlinck, Jean
Sihvola, Antti
Sivic, Husen
Sjoblom, Anna
Skoog, Anna
Skoog, Carl (child)
Skoog, Harald (child)
Skoog, Mabel (child)

Skoog, Margaret (child)
Skoog, William
Slabenoff, Petco
Smiljanic, Mile
Sohole, Peter
Solvang, Lena Jacobsen
Sop, Jules
Staneff, Ivan
Stoyehoff, Ilia
Stoytcho, Mihoff
Strandberg, Ida
Stranden, Jules
Strilic, Ivan
Strom, Selma (child)
Svensen, Olaf
Svensson, Johan
Svensson, Coverin
Syntakoff, Stanko
Tikkanen, Juho
Todoroff, Lalio
Tonglin, Gunner
Turcin, Stefan
Turgo, Anna
Twekula, Hedwig
Uzelas, Jovo
Van Impe, Catharine (child)
Van Impe, Jacob
Van Impe, Rosalie
Van der Planke, Augusta Vander
Van der Planke, Emilie Vander
Van der Planke, Jules Vander
Van der Planke, Leon Vander
Van der Steen, Leo
Van de Velde, Joseph
Van de Walle, Nestor
Vereruysse, Victor
Vook, Janko
Waelens, Achille
Wende, Olof Edvin
Wennerstrom, August
Wenzel, Zinhart
Westrom, Huld A. A.
Widegrin, Charles
Wiklund, Karl F.
Wiklund, Jacob A.
Wirz, Albert
Wittenrongel, Camille
Zievens, René
Zimmermann, Leo

Embarked at Cherbourg

Assaf, Marian
Attala, Malake
Baclini, Latifa
Baclini, Maria
Baclini, Eugene
Baclini, Helene
Badt, Mohamed
Banoura, Ayout
Barbara, Catherine
Barbara, Saude
Betros, Tannous
Boulos, Hanna
Boulos, Sultani
Boulos, Nourelain
Boulos, Akar (child)
Caram, Joseph
Caram, Maria
Chehab, Emir Farres
Chronopoulos, Apostolos

Chronopoulos, Demetrios
Dibo, Elias
Drazenovie, Josip
Elias, Joseph
Elias, Joseph
Fabini, Leeni
Fat-ma, Mustmani
Gerios, Assaf
Gerios, Youssef
Gerios, Youssef
Gheorgheff, Stanio
Hanna, Mansour
Jean Nassr, Saade
Johann, Markim
Joseph, Mary
Karun, Franz
Karun, Anna (child)
Kassan, M. Housseing
Kassein, Hassef
Kassem, Fared
Khalil, Betros
Khalil, Zahic
Kraeff, Thodor
Lemberopoulos, Peter
Malinoff, Nicola
Meme, Hanna
Monbarek, Hanna
Moncarek, Omine
Moncarek, Gonios (child)
Moncarek, Halim (child)
Moussa, Mantoura
Naked, Said
Naked, Waika
Naked, Maria
Nasr, Mustafa
Nichan, Krokorian
Nicola, Jamila
Nicola, Elias (child)
Novel, Mansouer
Orsen, Sirayanian
Ortin, Zakarian
Peter, Catherine Joseph
Peter, Mike
Peter, Anna
Rafoul, Baccos
Raibid, Razi
Saad, Amin
Saad, Khalil
Samaan, Elias
Samaan, Hanna
Samaan, Youssef
Sarkis, Mardirosian
Sarkis, Lahowd
Seman, Betros (child)
Shabini, Georges
Shedid, Daher
Sleiman, Attalla
Stankovic, Jovan
Tannous, Thomas
Tannous, Daler
Tannous, Elias
Thomas, Charles
Thomas, Tamin
Thomas, Assad (infant)
Thomas, John
Tonfik, Nahli
Torfa, Assad
Useher, Baulner
Vagil, Adele Jane
Vartunian, David

Vassilios, Catavelas
Wazli, Yousif
Weller, Abi
Yalsevae, Ivan
Yasbeck, Antoni
Yasbeck, Celiney
Youssef, Brahim
Youssef, Hanne
Youssef, Maria (child)
Youssef, Georges (child)
Zabour, Tamini
Zabour, Hileni
Zarkarian, Maprieder

Embarked at Queenstown

Barry, Julia
Bourke, Catherine
Bourke, John
Bradley, Bridget
Buckley, Daniel
Buckley, Katherine
Burke, Jeremiah
Burke, Mary
Burns, Mary
Canavan, Mary
Cannavan, Pat
Carr, Ellen
Carr, Jeannie
Chartens, David
Colbert, Patrick
Conlin, Thos. H.
Connaghton, Michel
Connors, Pat
Conolly, Kate
Conolly, Kate
Daly, Marcella
Daly, Eugene
Devanoy, Margaret
Dewan, Frank
Dooley, Patrick
Doyle, Elin
Driscol, Bridget
Emmeth, Thomas
Farrell, James
Flynn, James
Flynn, John
Foley, Joseph
Foley, William
Fox, Patrick
Gallagher, Martin
Gilnagh, Kathy
Glynn, Mary
Hagardon, Kate
Hagarty, Nora
Hart, Henry
Healy, Nora
Hemming, Norah
Henery, Delia
Horgan, John
Jenymin, Annie
Kelly, James
Kelly, Annie K.
Kelly, Mary
Kennedy, John
Kerane, Andy
Kilgannon, Thomas
Kiernan, John
Kiernan, Phillip
Lane, Patrick
Lemon, Denis

Lemon, Mary
Linehan, Michel
Madigan, Maggie
Mahon, Delia
Mangan, Mary
Mannion, Margareth
McCarthy, Katie
McCormack, Thomas
McCoy, Agnes
McCoy, Alice
McCoy, Bernard
McDermott, Delia
McElroy, Michael
McGovern, Mary
McGowan, Katherine
McGowan, Annie
McMahon, Martin
Mechan, John
Meeklave, Ellie
Moran, James
Moran, Bertha
Morgan, Daniel J.
Morrow, Thomas
Mullens, Katie
Mulvihill, Bertha
Murphy, Norah
Murphy, Mary
Murphy, Kate
Naughton, Hannah
Nemagh, Robert
O'Brien, Denis
O'Brien, Thomas
O'Brien, Hannah
O'Connell, Pat D.
O'Connor, Maurice
O'Connor, Pat
O'Donaghue, Bert
O'Dwyer, Nellie
O'Keefe, Pat
O'Leary, Norah
O'Neill, Bridget
O'Sullivan, Bridget
Peters, Katie
Rice, Margaret
Rice, Albert (child)
Rice, George (child)
Rice, Eric (child)
Rice, Arthur (child)
Rice, Eugene (child)
Riordan, Hannah
Ryan, Patrick
Ryan, Edward
Sadlier, Matt
Scanlan, James
Shaughnesay, Pat
Shine, Ellen
Smyth, Julian
Tobin, Roger

Index